Memories *from the* Far

PAUL CALLAGHAN

G000123394

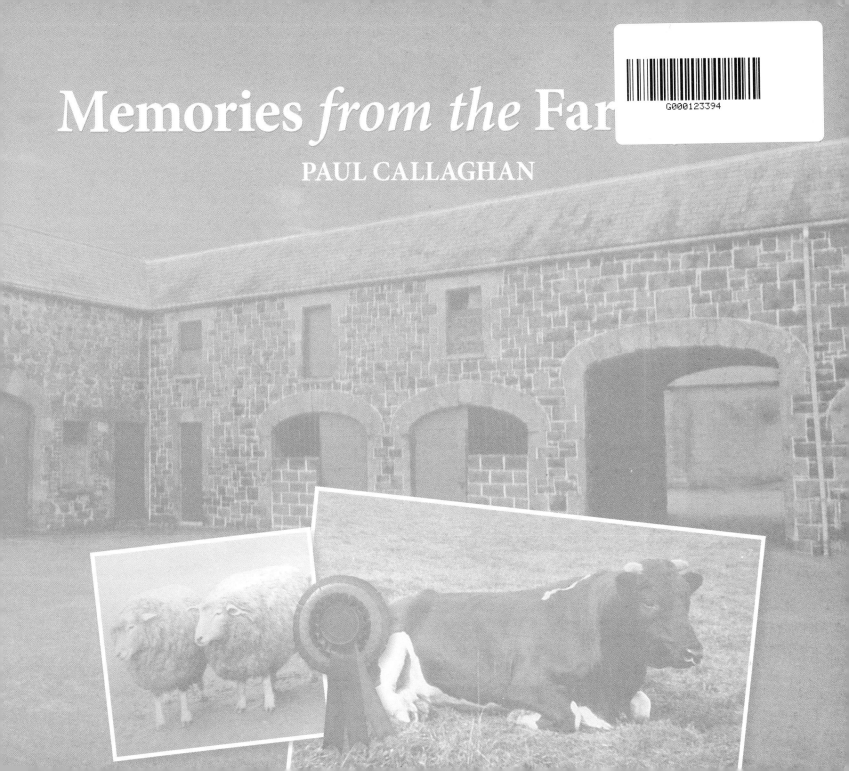

© Paul Callaghan

ISBN: 978 1 906578 63 3

First Edition

First Impression

Designed by April Sky Design

Printed by GPS Colour Graphics Ltd

All rights reserved. No part of this publication may be reproduced, stored in a retrieval system or transmitted in any form or by any means, electronic, internet, mechanical, photocopying, scanning, recording or otherwise, without the prior written permission of the copyright owners and publisher of this book.

Colourpoint

Colourpoint Books
Colourpoint House
Jubilee Business Park
Jubilee Road
Newtownards
County Down
Northern Ireland
BT23 4YH
Tel: 028 9182 6339
Fax: 028 9182 1900
E-mail: sales@colourpoint.co.uk
Web-site: www.colourpoint.co.uk

About the author

Paul Callaghan was born in Portadown, County Armagh where his interest in farming began at an early age. Much of his time outside school was spent on a small, mixed holding near his home. There he gained valuable experience and knowledge of older farming methods and traditional livestock breeds before moving on to assist on a modern pedigree pig unit. Later Paul studied meat technology and is a Graduate Member of the Institute of Meat. Since 2003 he has written a weekly column in *Farm Week* entitled 'Memories from the Farmyard'. He is married to Victoria and lives in Belfast.

Acknowledgements

The author would like to thank Mr Robert Irwin, Editor of *Farm Week* and Mr Hal Crowe, former Editor of *Farm Week* for their encouragement and support, and Victoria Callaghan for her assistance with and patience during the preparation of this book.

Picture credits

Mr A McConkey: cover image (background)

Brian Holgate: 45

British Breeds of Livestock, 5th Edition, Ministry of Agriculture and Fisheries (London, 1927): 87

Prof C Bryner Jones (ed), *Livestock of the Farm*, The Gresham Publishing Company: 91, 94, 95

Farmer and Stock-Breeder, Dorset House, Stamford Street: 85, 86

George Hammond: 41, 42

Henry Stephens, *Stephen's Book of the Farm*, William Blackwood and Sons, MCMIX: 4 (middle), 66, 68, 90

History of British Friesian Cattle, WE Baxter Ltd (Lewes, 1930) published for the British

Friesian Cattle Society: 33

Howard McCutcheon: 5 (left), 104, 105

Irish Farmers Gazette (1854): 15

Mrs M Thompson: 39

Paul Callaghan: 3, 4 (2nd from top, bottom), 5 (bottom), 6, 7, 8, 11, 12, 17, 31, 34, 38, 57, 60, 61 (right), 62, 63, 69, 71, 72, 76, 77, 79, 80, 81, 83, 92, 93, 106, 108, 110

Mr RA Bradstock, Free Town, Tarrington, Hereford: 75

Robert Wallace, *Farm Livestock of Great Britain*, Oliver and Boyd (Edinburgh, 1923): 21, 30, 61 (left), 97

Sam Hall: 27, 28

Sam Wilson: 47, 49, 52, 55

Foreword

I enjoy reading Paul Callaghan's weekly editorials in *Farm Week* where he brings an insightful perspective on the history and development of Northern Ireland's agri-industries. So it was with some anticipation that I looked forward to this opportunity to review his new book, *Memories from the Farmyard*. I must say that I was not disappointed and that I very much enjoyed Paul's book.

Memories from the Farmyard charts the development of all the principal cattle breeds in Northern Ireland across a period of just over one hundred years, and does so with remarkable flair and insight. The book reveals not just the methods used to create today's great breeds, but also gives flesh to many of the cattle breeders and characters from the past, whose vision and drive contributed so much to today's farming landscape. On a personal note, I have to say that as a retired dairy farmer, I have an innate interest in the history of cattle breeding in Northern Ireland. I was therefore particularly interested in the reference to the 'Ravenhill' herd, from which we bought our first Friesian bull in the late 1940s.

This book is a valuable record of how farming has evolved in this part of Ireland from the mid-nineteenth century and I am sure anyone with an interest in the history of farming and cattle breeding here will find it interesting reading. Moreover, I consider it to be a document that is well worth having on the bookshelf for future generations.

I have always believed that it is important to promote our farming industry to the consumer. *Memories from the Farmyard* is an important tool in this process. It shows the effort, work and vision that have gone into farming, particularly cattle breeding, over the generations. This legacy from our forefathers has helped to position us where we are today, with one of the most efficient dairy industries in Western Europe, possibly the world.

Robin Morrow
President
Royal Ulster Agricultural Society

A British Friesian bull at Portadown show around 1974.

Contents

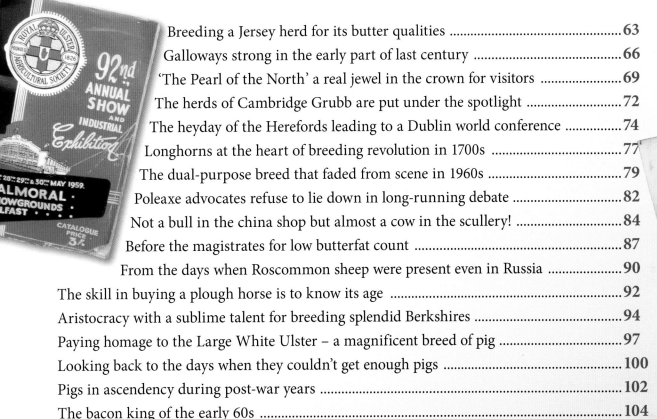

ADMISSION FREE. THE BULL WILL CHARGE YOU LATER

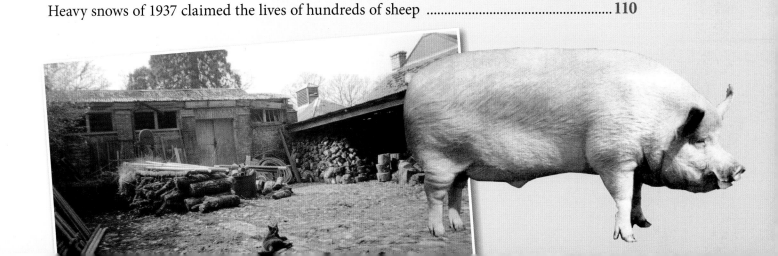

"In the year 1818, Sir Robert Bateson, Bart, of Belvoir Park, Belfast began to form a herd of Shorthorns, which he bred with success obtaining many prizes in later years at shows of the North-East Agricultural Society of Ireland".

Bateson – a pioneer of Shorthorn breeding

Turning the pages of James Sinclair's 'History of Shorthorn cattle' which was published over one hundred years ago, one reads the following: "In the year 1818, Sir Robert Bateson, Bart, of Belvoir Park, Belfast began to form a herd of Shorthorns, which he bred with success obtaining many prizes in later years at shows of the North-East Agricultural Society of Ireland".

View across the River Lagan towards the Morelands Meadow on the Belvoir Park Estate, where Sir Robert Bateson's Shorthorn cattle grazed between 1818 and 1863.

Perhaps, those *Farm Week* readers with an interest in Ulster's rich history of cattle breeding would be keen to learn more about Sir Robert Bateson given the fact that he was one of our earliest Shorthorn breeders. Before profiling this nobleman and his Shorthorn herd it may be of value to say that at the time he was founding his herd, our cattle were in great need of improvement.

According to the aforementioned author, the cattle of Ireland were "varied in colour, slow feeders and did not weigh more than five hundredweight when finished for the butcher at four or five-years-old". Old eighteenth and early nineteenth 'lost cow' descriptions printed in north of Ireland newspapers would most certainly validate James Sinclair's remarks.

For example in 1789 a beast strayed or was stolen from the townland of Ballymasca in the parish of Dundonald and was described as being "a red milch cow rising 11-years-old, about four hundredweight, broad made behind, a hole in one of her horns, no white but on her udder". Later, in 1790, another newspaper notice described a bullock and a cow that was stolen from the land of Reverend Doctor Torrens at Ballynaseereen, Aughnacloy. The bullock was "red without a white hair, straight up horns, lengthy and rising five" and the cow "brown and white, brown sides and white back and belly, speckled in the face about seven-years-old. Both weighed about four hundredweight".

So, the descriptions of these two cattle serve to underline the fact that our eighteenth century cattle were of poor types and as such when Sir Robert Bateson decided to import some Shorthorn cattle from the north of England, he was doing something hugely significant. Regarding this great breed history in Ireland, Sir Robert Bateson of Belvoir Park, Belfast was a trail-blazer.

In 1811 Robert Bateson married Catherine, the youngest daughter of Samuel Dickson Esq, Ballymaguille, County Limerick and around that time they took up residence at Belvoir. Their dwelling, once described as Belfast's finest and most historic mansion, was set in beautifully laid out grounds. Inside the stately building was an excellent library and the walls of most rooms were adorned with paintings, many of which in the early 1800s were reported as being "ancient".

Sir Robert Bateson (born 1780) was created a baronet in 1818. He was Deputy Lieutenant of County Down and represented Londonderry in Parliament between 1830 and 1842. Regarding the development of farming in Ulster, Sir Robert Bateson was to play an influential role. He was one of the leading subscribers to the North East Society, an agricultural association that was formed in 1826.

On Wednesday 9 April 1828, the Society held its Spring Show of cattle at May's Market, Belfast. It was subsequently reported that a "numerous assemblage of Noblemen and Gentleman" attended. A considerable number of "very fine animals" were exhibited, several of which had been bred and fed at Belvoir Park. These included some cows, heifers and fat oxen, all of which belonged to the Improved Shorthorn breed. That evening in Kern's Hotel, the Society held a dinner attended by around sixty people. These included Sir Robert Bateson, the Marquis of Donegal, Marquis of Downshire, Lord Ferrand, Lord Castlereagh, Sir James Anderson, Sir A Chichester and the Hon. Colonel Pakenham.

Later that year, on Thursday 18 September 1828, Sir Robert Bateson chaired the Moira and Maralin Branch of the North East Society's

SHORT-HORN BULL FOR SALE.

TO BE SOLD BY AUCTION, at DONALDSON'S REPOSITORY, DONEGALL-STREET, on WEDNESDAY, the 19th instant, at TWO o'clock, the Short-Horn BULL

WETHERELL.

He is a rich Roan, three years old, large and handsome; he was bred by Mr. SMITH, of Piersebridge, County Durham; is got by Emperor (A), dam by Harlsey (2,091); g. d. by Memnon (4,450); g. g. d. by Comus (1,861), &c., &c. Emperor, bred by Mr. SMITH, was got by Uranus (7,658), dam by Sol (2,655); g. d. by St. Leger (1,414); g. g. d. by Comus (1,861), &c., &c. Uranus was bred by Captain BARCLAY, of Ury.

WILLIAM RODGERS, Auctioneer.

Abbeylands, 14th April, 1848. 1259

annual dinner. It was held in Mr Murphy's New Market House, Moira. After the cloth was removed and several toasts were drunk, the evening was enlivened by the singing of several good songs. These festivities followed, what had been reported as, "an excellent show of cattle".

In the early nineteenth century, it was sometimes the practice for livestock owners to issue challenges to other breeders via the newspapers. In one such case, Sir Robert Bateson invited competition by announcing his intention to exhibit an Improved Shorthorn called 'Belvoir Park Favourite'. This bull had been imported by Sir Robert Bateson, having been bred by Mr Champion of Blythe, England. It was born in 1826 and was the son of a well-known heifer named 'Matchless'.

Finding his herd overstocked in the summer of 1829; on Tuesday 11 August Sir Robert held an auction at Belvoir of several cows and heifers that were in-calf to 'Belvoir Park Favourite'. Included in the draft was a beast called 'Young Snowdrop', described as being "one of the finest year-old bulls in Ireland." At the first cattle show of the Lurgan Farming Society, a farmer called Mr Bewick exhibited "a bull of superior description". His landlord, Sir Robert Bateson, had presented this animal to him, having paid sixty guineas for its mother.

In 1830 a cow called 'Nectarine', reared by a first rate breeder in the county of Durham was imported. She was brought to the Belvoir estate and subsequently served by a bull called 'Prince', a son of the previously mentioned 'Belvoir Park Favourite'. The cow gave birth to a male calf that was "light brown, blended with white and beautifully mottled". He grew into a bull with what was said to be "exquisite symmetry" and having been purchased by a farmer about half a mile from Moneymore, was let to cows at half a guinea each.

It was not just farmers but butchers benefitted from making use of the excellent stock raised at Belvoir Park by Sir Robert Bateson. In the 1840s competition among Belfast's sellers of meat was intense. Occasionally, several of them took out newspaper advertisements, inviting the public to come and view their beasts prior to slaughter. On 20 February 1845 a person going to the premises of John Gaffiken of 26, Corn-Market, would have been rewarded by the sight of "two of the primest heifers in the North of Ireland". Both these animals belonged to the Improved Short-horn breed and were estimated to weigh about 18 cwt each. According to old records, these animals had been bred by Sir Robert Bateson.

During the 1840s an imported bull called 'Wetherall' would have been seen grazing the pastures of the Belvoir Estate through which the River Lagan flows. This animal had been sold by William Rogers during April 1848 and was at Belvoir the following month. According to a newspaper notice published that month, gentlemen and farmers could have their cows served by 'Wetherall' at one guinea each, and half-a-guineas, respectively.

Sir Robert Bateson was present at a dinner held to mark the first show in Belfast, staged by the North East Agricultural Association of Ireland. The venue was Belfast's Music Hall, which was, given the occasion, well decorated with evergreens and impressive paintings. On one wall were the arms of the House of Downshire, produced in purple and gold on a background of rose-coloured silk. In front of this sat the Chairman, the Marquis of Downshire. When he called on Sir Robert Bateson to speak, the latter obliged and stated that, "the improvement in the farmers' classes is especially attributed to the dispersion of well-bred bulls throughout the country, even though they may not have good 'wives', they produced a much better stock".

A short time after making his last public appearance at a Church Education Society meeting in the Music Hall, Belfast, Sir Robert Bateson (aged 83) died on 21 April 1863. In his obituary it was recorded that, "in all relations of life he bore himself as a Christian gentleman and whether as landlord or friend of politician, his loss would be deplored."

Although the mansion at Belvoir Park was inhabited by various people in the years that followed, it was, by the mid-twentieth century, in a derelict state. Accordingly, it was demolished with explosives by the Territorial Army on Saturday 18 February 1961. Today, visitors to the Belvoir Estate can enjoy a walk through impressive woodland, get a great view across Belfast, and see the pastures that would have once grazed some of Ireland's earliest examples of improved Shorthorn cattle.

The battle between Bates and Booth for Shorthorn supremacy

Kirklevington farmer Ian Tate with his portrait of the 1839 prize Bates bull 'Duke of Northumberland'.

This Memorial OF THOMAS BATES, OF KIRKLEVINGTON, ONE OF THE MOST DISTINGUISHED BREEDERS OF SHORTHORN CATTLE, IS RAISED BY A FEW FRIENDS WHO APPRECIATE HIS LABOURS FOR THE IMPROVEMENT OF BRITISH STOCK, AND RESPECT HIS CHARACTER. BORN 21st JUNE 1776. DIED 26th JULY 1849.

Inscription on the gravestone of famous Shorthorn breeder Thomas Bates (1776–1849).

"Booth for the butcher and Bates for the pail" is an old Shorthorn adage which referred to the two types of Shorthorn favoured by the nineteenth century North of England cattle breeders Mr Thomas Booth and Mr Thomas Bates. Whereas the animals favoured by Mr Booth and his sons, Richard and John, were renowned for their flesh making capacity and breadth of back and loin, Mr Bates' animals were deep milkers, exhibiting what we would now call 'dairyness' and what the old writers would call 'gaiety of style'.

Following his death on 26 July 1746, Thomas Bates was buried in the graveyard of St Martin's Church, Kirklevington, near Yarm, Northumberland.

If one had wanted to see some excellent examples of Booth animals in the north of Ireland around 1880, a trip to Mr George Allen's farm at Unicarville, Larne, would have proved of benefit. Mr Allen was a 'Booth specialist' and for a number of years, annual on-farm bull sales were conducted at Unicarville by the well-known cattle salesman Mr Gavin Lowe of Dublin.

Staying in County Antrim during the 1880s, a bull rich in Bates blood called '15th Duke of Wellington', could have been seen in the hamlet of Magheramourne, five miles south of Larne, on the lands of James McGarel-Hogg, first Baron Magheramourne (3 May 1823–27 June 1890). This animal had, in fact, been bred by his father-in-law Edward Gordon Douglas-Pennant, first Baron Penrhyn (20 June 1800–31 March 1886).

When Sir J McGarel-Hogg, Bart took out a newspaper advertisement profiling the pedigree of his '15th Duke of Wellington' and announcing the intention to sell him at Robson's sale-yard, Belfast, in 1883, those readers with a good knowledge of Shorthorn would have perceived here was a 'milky' bull, full of the blood of Thomas Bates (1776–1849),who farmed at Kirklevington, near Yarm. The bull's dam was of the 'Waterloo' tribe sprung from a cow Bates had purchased in 1831. Another animal mentioned in '15th Duke of Wellington's pedigree was the first of Bates' prize bulls, which was called 'Duke of Northumberland'.

The first English Royal Show at Oxford took place during the year of 1839. While those who participate in Agricultural shows today have the luxury of motorised transport, back in the 1830s moving livestock over large distances was not so easy. It has been recorded that for Mr Bates the trip to and from Oxford took an incredible twenty-six days! For 'The Duke' it must have involved quite of lot of walking as it was reported that, over the trip, he lost three and a half hundredweight (175 kgs).

When describing another Bates bull, one of the old writers said it was "a grand dashing fellow, full of spirit, as fiery as a charger, with a noble head." Bates, it would seem, liked an animal with plenty of vim. On 23 June 1831, in a search for new blood, Thomas Bates rode over to a farm owned by John Stephenson of Wolviston. As the two men stood in the yard, possibly discussing Shorthorn ancestries, an inquisitive young bull shot his head over the half door of his box. Immediately seeing and sensing the animal's masculinity and power, Mr Bates is reputed to have shouted "Eureka!" Apparently, it took 'three strong men' to bring the animal to Kirklevington.

Thomas Bates was not behind the door when it came to espousing the excellence of his cattle and skill as a breeder. When a painter asked permission to commit their images to canvas he was told, "I do not expect any artist can do them justice. They must be seen and the more they are examined, the more their excellence will appear to the true connoisseur". Continuing on the theme, Mr Bates added that there were few true judges stating, "Hundreds of men may be found to make a Prime Minister for one fit to judge the real merits of a Shorthorn"!

Mr Bates also is quoted as saying that "By extra feeding the worst of animals may be forced to gains the applause of incompetent judges" and that he "hardly knew of any breeder who, if his livelihood depended on his own skill and knowledge would not die of want". This last wonderfully bold comment went some way to explaining why in his book *Types and Breeds of Farm Animal*, Professor Charles S Plumb stated Bates "did not make friends among the breeders".

We could extend Professor Plumb's statement with the words "or their wives". When the legendary County Durham Shorthorn breeder Charles Colling (1749–1836) held his dispersal sale in 1810, Thomas Bates secured a cow which was to become the foundress of his famous Duchess tribe. She'd been knocked down at humble money and when Mrs Charles Colling (nee Colpitts) heard of Mr Bates' determination to buy the animal she said if she'd known his enthusiasm, he'd have been "run to the uttermost farthing".

Of course Thomas Bates had a great belief in his abilities as a cattle breeder and although some people may have thought him somewhat egotistical he was, most certainly, a man to be admired. Ridiculed on occasions by the foolish, Thomas Bates was admired by those with discernment. According to the author James Sinclair (*History of the Shorthorn*) Thomas Bates' character through his Christian faith, brought him the respect of many people. In all his affairs Mr Bates followed an undeviating course of moral conduct.

Whilst Thomas Bates may have come across as having little time for his peers, this was most certainly not the case when it came to his cattle. Often he would have been seen speaking gently to them and stroking their heads. The cattle enjoyed this and if Bates gave too much attention to one beast, its neighbour would give him a poke with her horn. When it came to helping his stockman bring the cows in for milking Mr Bates was little more than a nuisance as they'd gather round him.

"I wish you'd keep out of way", the employee would say to his employer, "you do more ill than good for they won't leave you and there's no driving them!" It should be noted, however, that Mr Bates' great affection for his cattle was not carried too far. When one failed to breed, it was not long before she'd be sent to the knife.

In addition to the aforementioned 'Waterloo' and 'Duchess' families, Mr Thomas Bates developed the 'Oxford' tribe, 'Cambridge Rose' tribe, 'Wild Eyes' tribe and 'Foggathorpe' tribe.

Purebred Shorthorn bull, the property of Sir J McGarel-Hogg, Bart, Magheramourne, County Antrim

15th Duke of Wellington. Born 14th February 1880 by 42nd **Duke of Oxford** (39281). Bred by Lord Penrhyn of Penrhyn Castle, North Wales. **Dam Waterloo 45th** by **20th Grand Duke 11th** (218490), ggd **Waterloo 17th** by Red Knight (11976), gggd **Waterloo 14th** by **Grand Duke** (10284), ggggd **Waterloo 13th** by **3rd Duke of Oxford** (9047), gggggd **Waterloo 9th** by **2nd Cleveland Lad** (3408), ggggggd **Waterloo 6th** by **Duke of Northumberland** (1940), gggggggd **Waterloo 3rd** (2377), ggggggggd **Waterloo Cow** by **Waterloo** (2816), gggggggggd by **Waterloo** (2816).

Following his death in July 1849, the herd of Thomas Bates was sold by auction on Thursday 9 May 1850. This historic sale, which took place at a time of depression, realised better prices than expected with the sixty-eight animals forward fetching an average £67 0s. 6d.

Today in the graveyard of St Martin's church, Kirklevington, near Yarm in the County of Yorkshire, is a memorial stone to Mr Thomas Bates, "one of the most distinguished breeders of Shorthorn cattle".

The other half of the battle for Shorthorn supremacy

Last week we reminded you of the old adage "Booth for the butcher and Bates for the pail" – which referred to the two types of Shorthorn favoured by the nineteenth century North of England cattle breeders Thomas Bates and Thomas Booth.

Mr Bates, the subject of the previous 'Memories from the Farmyard', liked his animals to be sleek, deep milkers. Mr Booth preferred his cattle to carry a wealth of flesh on big, broad frames.

Ireland was once described as being awash with the blood of Booth so it may be good this week to look at how he and his sons John and Richard Booth fared. These gentlemen raised prize beasts on their North Yorkshire farms with such success that the names of the 'Killerby', 'Studley' and 'Warlaby' stand out in the Shorthorn breed's remarkable heritage. Although there were actually three generations of Booths involved in the Shorthorn business, our focus for this week will be on the founder's son, Mr Richard Booth (c.1788–1864). He first lived at Studley but later moved to Warlaby, near Northhalleron, Yorkshire.

Before the days of photography, farming correspondents wanting to convey the merits of a particular beast to their readers could do so in two ways. They could commission an artist to draw a representation of it, or they could use descriptive language to convey a mental picture. When the old Shorthorn historian Mr William Carr (*History of the Killerby, Studley, and Warlaby Shorthorns*, Ridgway 1867) wrote about Warlaby, he poetically singled out a cow called 'Charity' with the following words:

"Of 'Charity', who so long graced the Warlaby pastures, it is sufficient to say that she was the personification of all that is beautiful in Shorthorn shape. Such was her regularity of form that a straight wand laid along her side longitudinally, from the lower flank to the forearm, and from the hips to the upper part of the shoulder-blades, touched at almost every point. She was thrice decked with the white rosette at the Royal and thrice at the Yorkshire meetings".

A grandson of 'Charity', named Windsor, was held in such high esteem by Shorthorn fanciers in the 1850s that they linked his name with the famous bull that sold in 1810 for 1,000 gns. Windsor was, they concluded, "Comet of modern times". The Shorthorn historian James Sinclair relayed that Richard's Booth's bull 'Windsor' was shown ten times and won on nine occasions.

When the Great National Cattle Show of the Royal Agricultural Improvement Society of Ireland took place in the ancient city of Armagh on the 9, 10 and 11 August 1854, there were expectations for an event that truly lived up to its name. Not only would there be Devon, West Highland and Polled Angus cattle, but the entries for Shorthorns had doubled.

And so it was that on the Wednesday of show week, at six o'clock in the morning, the judges entered the showground on The Mall and commenced their duties. Eight hours later, having completed their deliberations they withdrew and the gates of the show-yard were thrown open and the public admitted at a cost of half-a-crown. Their Excellencies the Earl and Countess of St Germans were present as was the Earl and Countess of Clancarty, Lord Dunlo, Lord Talbot de Malahide, Lord Lurgan, Lord Claude Hamilton, Lord Dungannon, Sir Robert Bateson and many other distinguished personages.

The event took place under a bright sky and although the sun shone there were a few refreshing showers. As those in attendance moved round all the exhibits, they did so to the sounds of the 2nd Dragoon Guards band. A wonderful jet of water, rising some seventeen

Prize Shorthorned bull 'Windsor', calved 1 October 1851, got by 'Crown Prince', dam 'Plumb Blossom' by 'Buckingham', was exhibited by Mr Richard Booth, Esq of Warlaby, Northallerton, England at the Armagh Show in 1854. The bull was awarded a first prize in his class and Council Gold Medal as best of the aged bulls.

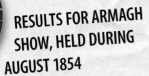

RESULTS FOR ARMAGH SHOW, HELD DURING AUGUST 1854

Class A Short-Horned Bull

Section 1: For best bull, calved on or after the 1st of January, 1849 and previous to 1st January 1852, 30 sovs. No 8: Richard Booth, Warlaby, Northallerton, England; short-horned bull Windsor, calved October 1, 1851; bred by exhibitor, got by Crown Prince; dam, Plum Blossom, by Buckingham; gd, Hawthorn Blossom, by Leonard.

For the second best ditto, 10 sovs No 1: Sir Arthur B Brooke, Bart, Colebrooke Park, Brooke-borough, short-horned bull, Rate-in-aid; calved January 27, 1949; bred by Richard Chaloner, Kingsfort, Moynalty; got by Hamlet (8126), dam Fanny the Second, by the General (3871), gd, Fanny the First; by Prince Ernest (7366), ggd, Flora, by Prince Arthur, own brother to Noble.

No 4: James G Wood, Castle Grove, Ballymaleel, Strabane, commended; Short-horned white bull, Comet, calved April 18, 1851, bred by Henry Ambler, of Watkinson Hall, Halifax; got by Broker (9993); dam, Fair Frances, by Sir Thomas Fairfax (5196); gd, Feldom, by Young Colling (1843).

No 5: Lord Dufferin and Clandeboye, Clandeboye, Holywood, commended; Short-horned bull, Kossuth, calved July 24, 1851; bred by exhibitor; got by King William (10379); dam, Flora the Third, by Petrarch (7329); gd. Lady Francis by Sultan Selim (2710), ggd, Francis, by Sir Frances B (1443), gggd, by Wellington (683); gggggd Flora of Pitcorthie by North Star (458), own brother to Comet (155).

No 10: Anthony Babington, Creevagh, Londonderry, commended, short-horned bull Sir Hugh, calved May 23, 1850, bred by exhibitor; got by Lord John, the property of Robert Holmes of Waterstown; dam, Betty Martin, by Sir John Sinclair; dam Young Miss Jones by Eclipse.

No 11: JW Maxwell, Finnebrogue, Downpatrick, commended, short-horned bull, Billy the Beau, calved June 13, 1850, bred by exhibitor; got by Beau Bill (9940); dam, Diadem, by Solway (7530), gd, Dew-drop, by Halley's Comet (2088); ggd, Diana, by Brutus (1750).

feet into the air, formed the centrepiece to the showground and presented a tremendous spectacle.

Before focussing on livestock, it may be of interest to list some of the implements on show such as Thomas Edgar's (Portadown) one-horse portable steam-engine, the highly-finished ploughs, grubbers and horse-hoes of Robert Gray, Police Place, Belfast, and the beautifully made plunge-churns, milk-coolers and steaming apparatus exhibited by local man William Hamill.

There was also sheep-dipping apparatus. While the latter item would undoubtedly have been of great interest to flock owners, they would have gleaned much on the stand of Joseph Long, of Meriton's Wharf, London. Mr Long was actually in attendance himself, on hand to espouse the great value of his medicines and lotions for treating such ailments as scab or foot-rot in sheep.

Those visitors to the 1854 Armagh Show with an interest in cattle were in for a treat. Exhibits included a lovely red Devon heifer, exhibited by Roger Hill, Narrow Water, Warrenpoint, and a Sussex heifer called 'Miss Rix' born 4 April 1851, from the Earl of Caledon, Caledon Hall. This animal had actually been imported from Sussex.

Native breeds included two Kerry heifers exhibited by John L Gaussen, MD, Crumlin, County Antrim, but what of the Scottish breeds? Sir Frederick W Heygate of Bellarena, Limavady, brought some of his West Highland's to the event, while a Galloway cow called 'Maid of Galloway' (calved March 1846 and a grand-daughter of 'Old Maid of Galloway') was entered by Mrs Gage of Bellarena, Limavady.

There were two more Scottish breeds at Armagh that year; Alexander Hutcheson, Gosford Farm, Markethill, successfully paraded an Ayrshire cow. The Polled Angus variety was also well-represented in the form of a bull called 'Monck' (calved 22 February 1852), which had been bred by the Scottish nobleman Sir James Carnegie sixth Bart (1827–1905).

But now to the Shorthorn entries which, as previously stated, had doubled for that year's 1854 Armagh Show. Local breeders were very much to the fore in the prize stakes and these included Samuel Orr of Flowerfield, Coleraine; Anthony Babington, Creevagh, Londonderry; Jonathon Richardson, Glenmore, Lisburn; Sir Arthur B Brooke, Bart, Colebrooke Park, Brooke-borough; JW Maxwell, Finnebrough, Downpatrick; Charles James Knox, Jackson Hall, Coleraine; James G Wook, Castle Grove, Ballymaleel Strabane; and finally Lord Dufferin and Clandeboye of Clandeboye, near Bangor, who showed a bull called 'Kossuth' (calved 24 July 1851).

Sir Arthur Brinsley Brooke (d.21 November 1854) of Colebrooke Park also showed a good Shorthorn bull in the form of a five-year-old called 'Rate-in-aid'. "Good" but not good enough the take the top prize. This didn't go to a local gentleman or farmer for that matter. Rather, it was awarded to none other than the aforementioned Shorthorn celebrity Mr Richard Booth of Warlaby, Yorkshire; with his nine-time Champion bull 'Windsor', the portrait of which is featured on the previous page.

When white Shorthorns were frowned on as lacking fertility and hardiness

This fine 'blue-grey' bullock was photographed in Scotland.

Commenting on a sale of pedigree Shorthorn cattle that took place in Belfast on Tuesday 26 March 1889, an agricultural reporter stated, "it was strange that all the white bulls were bought for Scotland". Although almost one hundred and twenty years have passed since the event, today it is possible to forward a hypothesis regarding the departure of those white-coated beasts. Before doing so, however, it may be interesting to profile the auction.

The first annual sale for pedigree Shorthorn bulls at Robson's took place in 1878 and so this one, in 1889, was the twelfth. During the mid to late nineteenth century a growing number of farmers in the north of Ireland turned to the Shorthorn breed (nicknamed 'The Great Improver') to raise the quality of the native cattle. This mushrooming interest was reflected in the fact that the number of entries for the 1889 sale exceeded all those that had come before. Around ninety pedigree bulls, principally yearlings, were due to come under the hammer. On the day of

PRICES OBTAINED AT SALE OF BULLS, ROBSON'S MART, BELFAST ON TUESDAY 26 MARCH 1889

The property of George Dickson, The Nurseries, Newtownards, Bendigo, red, 8 months, Mr Branagh, Belfast, **7 gns.**

The property of Mr AS Crawford, Crawfordsburn, Co Down, Bruce, rich roan, b.3/5/1888, Mr Waring, Moira, **12 ½ gns.**

The property of Mr Joseph B Wiley, Ballycushan, Templepatrick, Forest King, red and white, b.2/4/1888, Mr Casement, Ballycastle, **17 gns.**

The property of Wm. Charley, Seymour Hill, Dunmurry, Lord George, red and white, c.31/1/1888, Lord Deramore, Belvoir Park, Belfast, 30 gns.

The Property of Mr Anketell Moutray, Favour Royal, Aughnacloy, Lord Fauntleroy, white, c.25/1/1888, Mr Dunlop, Maybole, **25 gns**.

The property of Mr Samuel Jordan, Hillsborough, Alexander, rich roan, b.2/4/1888, Rev Mr Pakenham, Langford Lodge, **35 gns.**

The property of Mr Francis Quaile, Strangford, Riby King, roan, b.17/3/1888, Mr Shellington, Belfast, **19 gns.**

The property of Mr William Ash Gaussen, Ballyronan House, Magherafelt, Jupiter 2nd, red and white, c.2/3/1888, Mr John Gordon, Banbridge, **17 gns.**

The property of Mr Benjamin Dickson, Gilford House, Gilford, Lord Hartington, roan, b.6/2/1888, Mr Brown, Downpatrick, **25 gns.**

The property of Mr Henry Tohall, Moy, Co. Tyrone, Roe, red and white, calved 28/2/1888, Mr Shanks, Holywood, **10 ½ gns.**

The property of Mr Richard Cochran, Springhill, Quigley's Point, Londonderry, Glacier, white, b.12/7/1887, Mr Ralson, Stranraer, **17 gns.**

The property of Mr Joseph Carson, Ballydawley, Coagh, Northern Duke, dark red and little white, b.3/11/1888, Mr McMaster, Stranraer, **18 ½ gns.**

The property of Mr Michael King, Strangemore, Londonderry, Camnish, roan, b.6/3/1888, Mr McCaw, Dromore, **20 gns.**

The property of Mr Joseph Beck, Lissara House, Crossgar, Co Down, Champion, red and white, b.30/6/1887, Mr Potts, Whiteabbey, **12 gns.**

The property of Mr RP Maxwell, DL, Finnebrogue, Downpatrick, Goldheaver, red, b.26/3/1888, Mr Allen, Mountpanther, **9 gns.**

The property of Mr William J Oakman, Ashvale, Upper Ballinderry, Lurgan, Alphonso, dark red, b.23/2/1888, Mr Martin, Rathfriland, **15 ¼ gns**.

The property of Mr Joseph Atkinson, DL, Crowhill, Loughgall, Samuel, roan, b.3/1/1888, Mr Morton, Banbridge, **21 gns.**

The property of Rev. Hugh Hastings, Magheragall, Lisburn, Irish Hero, red and white, b. 1/1/1888, Mr Fair, Derriaghey, **14 gns.**

---

sale, Robson's well-known auctioneer Mr James Morton mounted the rostrum at twelve noon.

Having disposed of five miscellaneous Lots, it was time for Mr Morton to draw bids on a team of bulls that had come from the Castle Ward Estate, situated on the shores of Strangford Lough in Downpatrick. They were owned by Henry Ward, Fifth Viscount Bangor (1828–1911) and were all sired by 'Earl of Arundel'. The names of Viscount Bangor's yearlings were 'Claret' (born 1 January 1888), 'Diamond' (born 21 March 1888) and 'Admiral' (born 12 February 1888). The first two beasts were red roans but the third, unfortunately, was white.

At the time this sale was taking place, there was something of a resistance when it came to white Shorthorns, owing to the belief that they lacked the fertility and robust constitution of the reds, roans or red and whites. When a valuable cow in the herd dropped a calf with a pure white or predominantly white coat breeders could be disappointed because, when reared up and presented for sale as a breeding animal, it would fetch less money.

The aspersion that 'whites' lacked vigour was unproven and as such was challenged from time to time by farming correspondents. Writing in *Irish Farming World* over one hundred years

ago, one such contributor mounted a challenge by drawing attention to those hardy herds of white cattle kept at on the Chillingham Estate in Northumberland and Cadzow Park, in Lanarkshire. Both these groups had sprung from those white beasts that had been enclosed at the respective locations, back in the thirteenth century. Since that time, with minimal interference by their owners and with little or no introduction of outside blood, both groups of white cattle had prevailed through sun, rain and storm.

The aforementioned writer in that *Irish Farming World* newspaper of 1901 also drew attention to another robust white breed called the 'Charolaise' which was, they added, "owned by the noble Gaul". This white French breed and those white herds being out-wintered in British parks, tested the mistaken belief that pale coat coloured cattle were delicate. Despite all this, however, the notion was perpetuated for some time and at the aforementioned sale in Belfast during 1889, it reduced the overall prices for those breeders who had 'a white one' in the team.

Viscount Bangor's roan entries 'Claret' and 'Diamond' sold well realising 31 guineas and 28 guineas paid by Mr Houston, Orangefield, Belfast and Sir Richard Wallace, Lisburn, respectively. The white bull 'Admiral', however, only managed to fetch 25 guineas. According to a sale report "the average price of Viscount Bangor's exhibits was excellent but unfortunately Lot 7 (Admiral) was white". He was purchased by Mr Callender of Newtownstewart, Ayrshire, Scotland.

At this juncture, we can offer a possible reason as to why the white bulls at the Robson's Belfast sale in 1889 went to Scotland. Perhaps, these animals were taken over to sire 'Blue-grey' calves which were the result from mating a Shorthorn bull of the white strain with black Galloway cows. The offspring's coat was made up of an intermingling of its mother's black hairs with its father's white hairs, giving the distinctive blue-grey or slate effect. One commentator stated that the sight of young blue-grey calves skipping round their Galloway dams on a fine spring morning was one of the finest sights in all farming.

As the twentieth century unfolded, the demand for 'blue-grey' stores continued, especially in those border counties of Scotland and England. Following the sales of stores, it was the practice to put an old white bull into the ring. Eventually the demand for such sires justified the staging of auctions dedicated to the sale those Shorthorn bulls of the white strain.

Following a public meeting held in the Greenhead Hotel, Northumberland on 12 March 1962 the Whitebred Shorthorn Association was formed. A herd book was duly published and it contained entries of 2,310 males and 506 females, from around 134 herds.

In 1966 the official Whitebred Shorthorn sales were moved to Carlisle where they continue to be held each spring and autumn. Some *Farm Week* readers may consider attending this sale with the intention of buying a good white bull. They should note, however, that unlike Viscount Bangor's 'Admiral' at the Belfast Sale in 1889, he'll not be knocked down at 25 guineas!

Overcoming prejudice against white cows at turn of the century

At a time when some farmers were reluctant to breed from white Shorthorn females on the grounds that they had reduced fertility and hardiness, Mr Anketell Moutray of the Favour Royal Estate in County Tyrone, showed no such prejudice. Around the turn of the twentieth century a fine collection of 'white' cows could have been seen in his long-established herd. These included 'Princess Amy' (calved 20 March 1900), 'Cosy' (born October 11th, 1897), 'Luxury 11th' (28 June 1894), 'Snowflake' (25 September 1899) and 'Aileen' (calved 9 January 1899).

According to old Favour Royal estate records, this last cow was a full sister to Rev Mr Hall's bull 'White Bear' and had a pedigree which could be traced back to a cow owned by the famous Mr Thomas Bates (1776–1849) of Kirklevington in Yorkshire. This gentleman specialised in breeding cows with deep milking powers and had earned a worldwide reputation. Acknowledging Mr Bates' tremendous flair as a cattle breeder one of the old writers said that, "his eye had the light of decided genius".

Some of the noted cow families established by Thomas Bates included 'Duchess' (descended from a beast by 'Daisy Bull' and purchased by Bates for 100 guineas in 1804), 'Wild Eyes' (descended from a

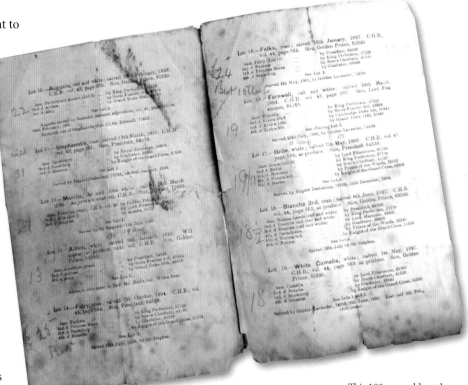

This 109-year-old catalogue provided potential buyers with details on the various Lots at the Favour Royal Shorthorn her sale on Wednesday 12 September 1900.

This fine white Shorthorn cow was photographed around 100 years ago.

roan heifer bought by Bates in 1832 for £3), 'Foggathorpe' (descended from Charles Collings 'White Bull'), 'Oxford' (descended from one of Bates' prize-winners at Royal Oxford Show in 1839) and 'Waterloo' (descended from a cow bought by Bates in 1831).

When Anketell Moutray (1844–1927) was introducing a new stock sire to his 6,000 acres County Tyrone Estate in the summer of 1903, he selected a descendant of the 'Waterloo' cow in the form of a young bull called 'Victorious' that had been born in August 1902. The bull's pedigree incorporated the best of English and Scottish, having as a grandsire the famous Scotch bull 'Bruiach', and a dam from a noted old English tribe.

Mr Moutray's young bull 'Victorious' had been born in August 1902 on the farm of Mr Peter of Browns Hill Farm, Berkeley, Gloucestershire. Reporting on this purchase the *Irish Farmers' Gazette* newspaper, dated 4 July 1903, stated that the bull "should prove a useful acquisition to a herd that had long ranked as one of the leading nurseries of the red, white and roans in the Northern Provinces."

Back in the 1840s, when the Favour Royal Estate was owned by Rev Moutray, some Shorthorn bulls were introduced for crossing with dairy cows. The progeny for this union did well and served to fuel Rev Moutray's interest in the Shorthorn breed. By the 1850s, he had founded his own pedigree herd having purchased some prizewinning heifers.

By the 1870s numbers had been sufficiently built up to stage two drafts sales, the first in 1872 and the second in 1878. Just as previously mentioned that there had been a good representation of white animals at the turn of the twentieth century, back in the 1870s the 'Favour Royal' herd included two such cows named 'Gertrude' (born 20 June 1872) and 'Governess 2nd' (born 9 September 1874).

WANTED LAND STEWARD

for Favour Royal Demesne.

Wages offered 20 shillings per week, with free house and 2 tons of coal per annum; if married man, grass for cow; single man, room in yard.

Not over 35 years.

Anketell Moutray, Favour Royal, Aughnacloy.

1905 recruitment notice for Favour Royal Estate

During this decade Rev Moutray brought his cattle up from Aughnacloy to Belfast to compete at the North East Shows which were held in the Cattle, Fax and Fruit Markets of East Belfast. On one particularly successful trip he won prizes with a bull called 'The Governor', a cow called 'Princess of Warlaby' and two heifers called 'Lady Gay' and 'Chaumontel'.

When Reverend John James Moutray, who was Rector in the County Tyrone parish of Errigal Keerogue, died, aged 84 years, on 20 June 1886, his County Tyrone Favour Royal Estate (approximately 6,700 acres) passed on to the aforementioned Mr Anketell Moutray DL. The on-farm draft sales of pedigree Shorthorn cattle continued and the first of the twentieth century took place on Wednesday 12 September 1900. It was conducted by Belfast auctioneer Mr John Robson and involved the sale of 32 head made up of pedigree bulls, cows and heifers. Potential customers could have made their way to the sale by taking the Clougher Valley light railway.

Trade on the day was brisk, especially for those females in-calf to stock bulls 'Golden Lavender' or 'Rupert Dashalong'. Lot 7,

'Countess' – a 2nd prize-winner at Belfast in 1893 made £32; Lot 11; 'Stephanotises' – due to calve down to Dashalong in October made £31; Lot 20, 'Charity' - had been running in a field with the Lavender bull – made £18½; Lot 15, 'Falka' – a breeder's beast – fetched £24, paid by Mr WE Best of The Cairn, Aghalee. Over the next few years she gave birth to six pedigree calves in Mr Best's herd, some of which were retained for breeding and others sold for respectable sums.

Two other farmers who also benefitted by investing in Favour Royal Shorthorns were Mr Benjamin Whann, Springdale Farm, Moneymore and Mr Thomas Campbell, Lisnamaghery, Clogher. When the Government of Northern Ireland Ministry of Agriculture issued the first volume of its 'Register of pure-bred dairy cattle' both these gentlemen had bought-in Favour Royal cows included in the 'Advanced Register'. Mr Whann's 'Dora' (white, calved 28 June 1898) gave 4,814 lbs (2184 kgs) at 3.9% and Mr Campbell's 'Princess of Orange' (roan, calved 1 July 1904) gave 8,001 lbs (3629 kgs) at 3.8%.

A year after the birth of the later animal, Mr Anketell Moutray had a position for a new Land Steward and therefore placed an advertisement in the *Irish Farmers' Gazette* newspaper. The wages were set at twenty shillings per week and the successful candidate would be given free accommodation, free grass for a cow and free coal. Although his advertisement discriminated against anyone born before 1870, all-in-all it presented a rather attractive package.

Foundation of noted Magheramourne Shorthorn herd on parade at Balmoral

Having exhibited some of his cattle at Balmoral show in 1935, it was reported that although Major HC Robinson of Magheramourne was a relative newcomer he should, in the near future, become a most successful competitor. This was attributed to the "magnificent foundation" with which he was establishing the 'Magheramourne' herd of pedigree Shorthorn cattle.

At this time Major Robinson's herd was in its infancy and thus, the animals he turned out at Balmoral were being used to form the bedrock on which it would be built. The Major's Balmoral team in 1935 comprised two roan females, 'Millbank Princess' and 'Augusta Hesba' and, a Scotch-type Shorthorn stock bull called 'Cruggleton Premier'.

Mr Alfred J Marshall of Bridgebank, Stranraer, had bred the latter beast and his herd was, at the time, the largest in Great Britain and Ireland. It comprised around 900 head, spread over five farms located on a narrow strip of land projecting into the sea, between Luce Bay and Wigtown Bay. Mr Marshall's cattle were registered under two prefixes 'Bridgebank' and 'Cruggleton'. According to the author Wing-Commander Marson (*The Scotch Shorthorn*) when the lists of names that could be attached to

the Bridgebank prefix had been exhausted, a start was made with 'Cruggleton'!

Before expanding on Major Robinson's early achievements at the Balmoral Show in 1935, it is worth outlining the development of the Scotch Shorthorn in general and the Cruggleton herd in particular.

When the 'Father' of the Shorthorn breed, Mr Charles Colling (d.1836) and his brother, Robert (d.1820), turned their attention to improving the cattle of County Durham and North Yorkshire they did so with great success. By selective breeding they created animals that were noted for both size and milking powers. Following the dispersal of Charles Colling's herd in 1810, two other breeders made full use of his bloodlines and earned themselves a place in breed history. Their names were Thomas Booth (d.1835) and Thomas Bates (d.1849).

While the former, Mr Booth (and his sons

Major HC Robinson of Magheramourne, County Antrim with some of his prizewinning Shorthorn cattle, photographed in the 1930s.

and grandsons) concentrated on creating cattle with a "wealth of flesh and quality and a deep broad frames on short legs" the latter breeder, Mr Bates, sought to develop a deep milking cow "of gay carriage, with a mellow flexible skin and much soft, mossy hair". This twin-track approach was later reflected in the old Shorthorn adage, "Booth for the butcher and Bates for the pail".

When the Shorthorn blood spread to Scotland, it was the Booth type beef animal that was promoted with great success, especially in the case of one breeder called Amos Cruickshank (d.1895) of Sittyton, Aberdeen. He placed more attention on type than pedigree and in time 'Sittyton' cattle were renowned for having hearty constitutions, good middles and frames, and a great capacity to fatten.

Following the dispersal of the Sittyton herd in 1889 the work to develop the 'Scotch Shorthorn' type continued and by the early to mid nineteenth century, many well known, distinguished prefixes had been established. These included 'Bapton', 'Calrossie', 'Cluny', 'Ermis', 'Gainford', 'Harviestoun', 'Millhills' and 'Uppermill', not forgetting the aforementioned Bridgebank and Cruggleton outfits owned by Mr Alfred J Marshall who bred our 1935 Balmoral Champion, 'Cruggleton Premier'.

Alfred J Marshall would have been introduced to the Scotch Shorthorn world at an early age, as his father, Matthew Marshall, was a well-known exporter. He organised the shipping of many valuable animals including one pedigree female sourced in Ulster. Her name was 'Princess May 3rd', born 28 February 1899 in Messrs JGW and WR Lyle's herd (commenced 1857 and dispersed 1901) at Donaghmore House, Donaghmore, County Tyrone. Having been purchased by Matthew Marshall, 'Princess May 3rd' was sold on, with her bull calf, to a Mr Mellish of Cape Town, South Africa.

During the first half of the twentieth century Bridgebank and Cruggleton cattle won prizes all over the world. In 1923 the herd supplied the champion bulls on two occasions at the great International Show in Chicago, while at a Buenos Aires event in 1925 'Bridgebank Masterstroke' was the sire of the reserve grand champion bull. 'Bridgebank Orpine' won at the Rand Show, Johannesburg in 1926 and the

following year 'Bridgebank Bayardo' lifted the Championships at both the Royal Dublin and Royal Ulster events.

In addition to being a successful breeder of Shorthorn cattle, Alfred J Marshall was a noted breeder of Clydesdale horses. The breed's 58th Stud Book stated that Bridgebank 'travelling' stallions were being extensively used throughout Great Britain and Ireland. These included 'Bridgebank Given' (21964) and 'Bridgebank Merryman' (21306), two Government Register horses, touring the districts of County Tyrone and County Armagh, respectively. The Studbook also gave details on twenty-one foals born on Mr Marshall's 'Bridgebank' holding during 1935.

That same year the previously mentioned Shorthorn breeder Major HC Robinson of Magheramourne paraded his stock bull, 'Cruggleton Premier', at the 1935 Balmoral Show. This animal, born 6 June 1932 was out of 'Lutwyche Pauline' and sired by 'Balcairn Colonel'.

Having settled his bull in a stall at Balmoral on the day before judging, Major Robinson may well have cast an eye over the opposition, as their owners busied themselves with curry comb and dandy brush. In the class for 'Bulls calved prior to 1933' the Major's roan bull would be up against two top class 'whites' in the form of Messrs FH and TT MacLean's (Glennane) 'Millhills Opal' and Mr RJ Linton's (Broughshane) 'Springmount Aide-de-camp'.

Judging of the Shorthorn classes for the Balmoral Show of 1935 got underway on Wednesday 29 May 1935. Placing the entries at that Silver Jubilee event were Mr Robert LP Duncan of Pitpointie, Auchterhouse, Angus and Senor Guillermo St J Peters, 499 Lavelle, Buenos Aries.

Although the number of Shorthorns being shipped across the Atlantic was falling during

the 1930s, it still was a lucrative trade and, having the services of an Argentinean judge at the 1935 Balmoral, was considered something of a coup. If this gentleman travelled home to give a good account of the Shorthorns he had seen, there could be benefits for Ulster breeders.

The task of ranking those Shorthorn cattle forward at the 1935 Balmoral was difficult because, according to one judge, "there were no plums". This gentleman meant that there were no particular beasts that clearly stood out from the others in their class. Hearing this remark a bystander may have taken this as a criticism of the cattle forward, however, it was to the contrary. When the judge in question said there were "no plums", he meant that they were in fact "all plums"!

Having decided that both 'Cruggleton Premier' and 'Millhills Opal' were up for the first and second tickets in their class, the Scottish and Argentinean judges could not agree on the order. A debate ensued and it was only after the intervention of a referee that Major Robinson's 'Cruggleton Premier' was placed at the top, going on to lift the Supreme Champion's gold medal and Robinson Cup for best bull.

It had been a great start for Major HC Robinson of Magheramourne and surely in the days that followed, he must have celebrated the 'wonderful foundation' of his Shorthorn herd. The 1935 Balmoral had been a great start and with the benefit of hindsight we can say that for Major Robinson and his Magheramourne Shorthorns, "there was so much more to come"!

See, the conquering hero comes!

Bedecked with rosettes, Shorthorn bull 'Magheramourne Masterkey', born 3 January 1935 and bred by Major HC Robinson of Magheramourne.

H aving been awarded the Championship for his stock bull at Balmoral in 1935, novice Shorthorn breeder Major HC Robinson of Magheramourne House, County Antrim, was reported as having got off to a great start. It was further acknowledged that, having laid such a good foundation for his pedigree herd, Major Robinson was likely to become a most successful competitor in the near future.

These words were both generous and optimistic, most unlike those experienced by a friend of mine who, around three decades ago, brought out the supreme champion at a particular pedigree show and sale. It was a proud moment and many people came over shook hands and congratulated him on the achievement.

One 'veteran' breeder, whose animal had been out of the running was, however, a little less magnanimous. Sidling up he muttered through clenched teeth, "Aye, it not too hard to get the top once but staying there, that's a different thing – there's only one way for you to go now!" It was rather a churlish remark but as it all turned out … stunningly accurate.

Whether Major Robinson incurred such a morose reaction from other exhibitors at the 1935 Balmoral is of little consequence. Studying old show reports for the years that followed we can confidently say that for Major Robinson's 'Magheramourne' herd, the only way was 'Up'.

At the time 'Cruggleton Premier', 'Millbank Princess' and 'Augusta Hesba' were being successfully paraded across the Balmoral turf on Wednesday 22 May 1935, a four-month-old red bull calf would have been seen knocking about the Magheramourne farm holding back home. Sired by 'Cruggleton Rock' and born in January that year out of the aforementioned

'Millbank Princess', this young calf had been given the name 'Magheramourne Masterkey'. Although the Major won many top awards during the 1930s with his cattle it was the 'Masterkey' bull that, one could say, 'put' most of the silver cups, tankards and medals on the sideboard within Magheramourne House.

Having reached 15-months-of-age 'Magheramourne Masterkey' made his Balmoral debut just two weeks or so after he had been awarded the Championship at the Royal Dublin Society's Spring Show of 1936. For 'Masterkey', Balmoral would not be as easily taken as Dublin, owing to the presence of another bull in his class called 'Broommount Pipe Major' bred by Mr Thomas Hayes of Broommount, Aghalee, County Antrim.

This animal, born 4 March 1935, was of Scotch Shorthorn type, his sire having been purchased at the Perth Bull Sales. The pedigrees of both 'Magheramourne Masterkey' and 'Broommount Pipe Major' were entered in Volume 82 of the *Coates's Herd Book* of the Shorthorn Society of Great Britain and Ireland. Being of similar age, type and colour, both bulls presented the judge at the 1936 Balmoral with something of a challenge.

They were closely matched but in the end the decision was taken to place 'Broommount Pipe Major' at the top of his class, with 'Magheramourne Masterkey' taking the second ticket. It was significant that both animals went on to be awarded the overall championship and reserve championship, respectively, in the Shorthorn Section.

Some time later, 'Broommount Pipe Major' was sold at Perth for 349 guineas and as such was out of contention when 'Magheramourne Masterkey' made his second appearance at Balmoral in 1937. This time Mr Kenneth P

MacGillivray of Kirkton, Bunchrew near Inverness, judged the Shorthorn classes.

Mr MacGillivray started his 'Kirkton' herd in 1914 and in addition to cattle his other farming enterprises centred on growing stock seed King Edward potatoes and a four hundred head flock of Cheviot ewes. Writing about Mr MacGillivray in a book titled *The Scotch Shorthorn*, Wing-Commander Marson stated that he had "few equals and no superior as a judge of pedigreed Shorthorn cattle."

It would seem that Mr MacGillivray was most impressed with the Shorthorn turn-out at our 1937 RUAS, especially those from the Magheramourne herd, which won practically all the cups. Speaking about Major Robinson's foundation cow 'Glennane Clipper 19th', Mr MacGillivray stated that she was the best seen in Belfast for a number of years, adding that he "considered her one of the best specimens of the Shorthorn breed, fit to win in any company". This cow won the Chichester-Clark Memorial Cup for best Shorthorn female.

'Magheramourne Charity' won the 'Victory' Cup for best yearling heifer and the Robson Memorial Cup for the best heifer not exceeding two years, bred and exhibited by a resident of Ireland. The 'Victory' Cup for best yearling bull went to 'Magheramourne Rampant' (red, calved 23 April 1936, s. 'Cruggleton Rock', d. 'Castledawson Lady Lucy') but what, *Farm Week* readers may ask, about the previously mentioned 'Magheramourne Masterkey'? How did he do? Well, at this his second time at Balmoral, 'Masterkey' took the Robinson Challenge Cup for Best Bull and the Gold Medal for the Supreme Shorthorn Champion.

When interviewed for a show report following his great achievements at the 1937 Balmoral Major Robinson said, "I have only

been in the game for four years and all the animals in the herd I either bred or bought myself without the assistance of anyone. It is therefore very gratifying that at the start I am on the right lines".

Describing Major Robinson's successful participation at the next Balmoral Spring Show of 1938 one reporter, using the title of a musical piece composed by George Frideric Handel (1685–1759), stated it was a case of "See The Conquering Hero Comes". With his "magnificent quartet" led by the 'Masterkey' bull, Major Robinson MP had swept the boards.

By this time 'Magheramourne Masterkey' had been a champion or reserve champion at Balmoral for three successive years and although not quite level enough on top or as full in the rump as some would have liked, most Shorthorn commentators were of the opinion that there was nothing in the showground to overstep his "low, thickset figure". Fashionable Shorthorns at this time were block-shaped and dumpy, with little light getting through underneath!

'Magheramourne Masterkey' made his final appearance at what turned out to the last pre-war Balmoral show, held during May 1939. King George VI's Land Stewart, Mr Alexander Ritchie MVO, who looked after the Royal herd of Scotch type Shorthorns at Windsor, was judge on this historic occasion. Impressed by 'Magheramourne Maskerkey's fine qualities, Mr Ritchie nominated him Supreme Champion. This was the bull's fourth appearance at the Balmoral Spring Show and his third successive year as overall Champion. With most of the other cups and trophies having been won by cattle owned by Major HC Robinson MP at this historic show, it was a case of "See the Conquering Hero Comes" – again!

> *"Looking back over old show reports it is clear that at the same time Sir Hugh's horses were being showered with high honours, his Shorthorns were experiencing the same on the cattle lawns."*

"We recently had the pleasure of visiting Mr HH Smiley's farm, Ardmore, Larne, County Antrim, were there has now been formed a select herd of pedigree Shorthorn cattle" wrote a correspondent for *Irish Farming World*, dated 27 April 1900. They added that "its composition clearly showed the judgement which has been exercised by the owner's agent, Mr James Coey".

Before examining the detailed description of Mr Smiley's 'Ardmore' herd of pedigree Shorthorn cattle, as recorded in the *Irish Farming World* newspaper published over 108 years ago, it may be useful to provide some background information, especially regarding the herd owner.

Hugh Houston Smiley, born 5 January 1841 in Islandmagee, was the son of Mr John Smiley, who conducted an extensive business as a corn merchant. Having been educated at Royal Belfast Academical Institution, Hugh Smiley joined his father's business, which was based in Larne. During these early years he developed a deep affection for this area and it was here that Hugh Smiley later set up his own home, in a picturesque residence known as 'Drumalis'.

Mr Hugh H Smiley married Miss Elizabeth Kerr of Gallowhill, Paisley, Scotland, the only child of Mr Peter Kerr who owned an extensive thread manufacturing company. In the course of time 'Gallowhill' also became one of Mr Smiley's homes and each year his time was split between this Scottish location and 'Drumalis' in Larne, County Antrim. In later years, owing to poor health, Mr Smiley found it necessary to winter abroad, usually in Egypt.

During those periods spent in Larne, Mr Smiley played an influential role in public life. Many charitable organisations and institutions benefited from his generosity, including the Countess of Dudley's Nursing Fund and the 'new' Royal Victoria Hospital. Mr Smiley also presented the Presbyterian Church with a

Sir Hugh Smiley's Shorthorn cow 'Lady Muriel'.

magnificent set of bells for its new assembly building in Fisherwick Place and in 1903 the 'Smiley Cottage Hospital' was opened. In August of that same year, a baronetcy was conferred on Mr Smiley.

Before profiling Sir Hugh Smiley's Shorthorn herd, it was worth mentioning that he owned a magnificent horse stud which was also visited in

Sir Hugh Smiley's Stock bull 'Candahar' 78522, born c.1897.

April 1900, by representatives of the newspaper *Irish Farming World*. At the time this 'Inver' Stud comprised forty horses, all kept in the most up-to-date accommodation. Two stallions that certainly made an impression on the visitors were the Thorough-bred 'Mascarille' and Hackney 'Gratian'.

The former stood at 16 hands and had been bred by Mr Leopold de Rothschild (1845–1917), the British banker of Ascott House, Ascott, Buckinghamshire. He was described as being "a finely-made horse showing both substance and quality with beautiful shoulders, a good back and powerful quarters." 'Mascarille' was registered as a 'Government Sire' and his progeny were reported as giving a good account

BALMORAL 1902

At the Royal Ulster Agricultural Show held in the Balmoral showgrounds during April 1902, Mr HH Smiley (later Sir Hugh Smiley) topped the twenty-four strong class for 'Shorthorn Bull, calved prior to 1900' with his Scottish-bred 'Candahar'. At the same event in the Horse section, Mr Smiley's thoroughbred Stallion 'Mascarville' also claimed first honours. Mr Smiley was vice-President of the Royal Ulster Agricultural Society.

DESCRIPTION of several pedigree Shorthorn cows breeding in Sir Hugh Smiley's Ardmore herd, Larne, as given by representatives of *Irish Farming World* in April 1900.

Stella, a deep, milking, heavy matron of great quality, bought at Clermont Park dispersal and is now just a dozen-years-old. She is light roan colour. Her rising two-year-old bull Seneschel is a dark roan of great depth by Lifeguard 7069.

Jenny Lind 24th, bred by Mr Duthie, Collynie, sired by the famous Scottish Archer 59893, a deep, thick, fleshy female of great frame that has a nice roan heifer calf at foot by Candahar.

Sunbeam 5th, bred by Mr John Cran at Keith and sired by Steady Lad 6640. In calf to Candahar.

Primrose 5th, in store condition bred by Mr Cran.

Lancaster 40th, grandly developed three-year-old heifer by Clarendon 68373.

Roan Butterfly, strong, light roan that is straight along the top with nicely laid tail and very good head.

Vain Fancy is a blocky well-proportioned Collynie-bred heifer.

Royal Welcome is another Collynie-bred heifer, two-years-old by Count Arthur 70194.

Lady Maud, a grand cow from Mr Marr, Uppermill.

Butterfly 29th, a four-year-old white cow displaying splendid frontage, well turned at ribs and great strength through the heart.

Lady 17th, first prize-winner at the Highland Show in Edinburgh and Royal Dublin Show. She has a good heifer calf by Peter the Great and possesses a beautiful head, great front and tine level top with a great wealth of flesh. Her shoulders leave nothing to be desired, her tail is splendidly laid in, while through the heart she is excellent, and likely to carry off a host of leading honours and the coming shows.

Granuaile 12th, six-year-old blocky symmetrical cow that carried off many honours for her breeder, the late Lord Caledon.

Myrtle 3rd, compactly built, rising three-year-old cow that show evidence of genuine merit and has a calf by Candahar.

for themselves at the Royal shows of Dublin and Belfast, besides being fine performers in the hunting field and natural jumpers.

The aforementioned Hackney sire, 'Gratian', was described by the visitors from *Irish Farming World* as being "a good dark bay with black points". He was by 'Old Time' (1863) which, at 15-years-of-age in June 1897, had carried a twelve stone man over a three mile grass track in a fraction less than ten minutes and thirty seconds.

Hackney Mares to be seen at Sir Hugh Smiley's 'Inver' Stud in April 1900, included 'Approval' 8647, a full sister to 'Applause' which sold for £1000, 'Mabel 2nd' 1236 described as being "15-years-old but in harness might pass as a 4-year-old", 'Queen Bess', a prize-winner at the prestigious London Hackney Show and the chestnut 'Merry May' 8247, winner of the Challenge Cups at both Belfast and Dublin.

Looking back over old show reports it is clear that at the same time Sir Hugh's horses were being showered with high honours, his Shorthorns were experiencing the same on the cattle lawns. The highly-coveted 'Queen Victoria Cup' competed for each year at Dublin, for instance, was won by Ardmore Shorthorns on three occasions.

When the aforementioned representatives from *Irish Farming World* came to Ardmore in April 1900, they were highly impressed with the cattle on the farm and their description of several of the cows is published alongside. The stock bull on the farm had been bred by Mr John Wilson of Pierriesmill, Scotland. This animal's name was 'Candahar' 7852 and he was described by the visitors as "a short-legged, strong-ribbed sire" with "just the stamp to sustain the reputation of the herd".

One of the cows to draw favourable comment from the visitors was 'Mable 3rd', also bred by Mr Wilson of Scotland. They stated, in their newspaper report, dated 27 April 1900, that one of her 'Candahar'-sired calves was also on the farm. Little did they know this red 12-week-old heifer was destined to make her own mark on Ireland's rich history of Shorthorn cattle breeding.

Having been given the name 'Myosotis', this heifer (born 6 January 1900) was reared up and sent out from Sir Hugh Smiley's Larne-based Ardmore herd, to join that owned by King Edward VII at the Royal Farm, Windsor. According to the Shorthorn Society's *Coates's Herd Book*, volume 55, part 2, one hundred years ago, on 24 January 1908, the Ulster-bred 'Myosotis' gave birth to a red heifer calf in the Royal herd at Windsor.

"Although down through the years a few farmers have dismissed the exhibitions of dairy cows as little more than a 'beauty' contest, the public milking trials offered a cow the chance to demonstrate her 'generosity' at the pail."

When Lincolnshire Red Shorthorns won the Balmoral milking trial

This pedigree Lincolnshire Red Shorthorn cow was bred in the famous Burton herd owned by Mr John Evens of Lincoln.

As we flit from Friesians to Ayrshires, then Shorthorns and back to Friesians in this column week after week, some *Farm Week* readers may be growing weary of it all and hope for something different. For these people this week we offer something fresh. The focus this week is not on any of the aforementioned breeds but rather on one that, perhaps, will not be so familiar – Lincolnshire Red Shorthorns.

At this juncture some readers may, on seeing the word 'Shorthorn' in this 'fresh' piece, experience a sense of déjà vu. But no! Although one word in this breed's name may be familiar, the 'Shorthorn' and the 'Lincolnshire Red Shorthorn' were two distinct breeds. Indeed, when the Articles of Association were being signed to form the 'Lincolnshire Red Shorthorn Association' in 1895, the Shorthorn Society sent a stiff letter objecting to the inclusion of the word 'Shorthorn' in the new body's name. Responding to the complaint, the Lincolnshire Red Shorthorn Association members stated that no-one had a monopoly on the word 'Shorthorn'.

If one had been taken on a tour of the farm of Gilford Castle Estate by owner Miss Katherine Carleton in 1906, some lovely milkers would have been seen in the byre, which was located in the corner of the yard. It's not hard to imagine entering the building through the wooden doors and hearing the clink of chains as cows turned their heads towards those making an entrance. There would have been some fine specimens belonging to the Shorthorn, Ayrshire and Jersey breeds because, as old

View across the farmyard at Gilford Castle towards the old byre, within which Miss Carleton's cows would have stood around 100 years ago.

records show, several of Miss Carleton's animals were 'show cows'.

Several animals with deep cherry red coats would also been seen down the line and these would have been pedigree Lincoln Red Shorthorns. Pointing them out, Miss Carleton may have been heard saying, "Those were bred by John Evens of the Burton herd". This remark would not have been lost on any visitor who was, shall we say, up-to-date with the 'who's who' of cattle breeding. Mr Evens (c.1835–1948) of Burton, near Lincoln, Lincolnshire owned one of the world's most successful dairy herds of Lincolnshire Red Shorthorns.

Farming in close proximity to the inhabitants of Lincoln, Mr Evens recognised the opportunity to sell milk and as such commenced dairy farming in 1885 with the native cows. At the beginning of that century, these had been improved following the introduction of Red Shorthorn bulls from the north of England. Right from the start, Mr Evens started weighing the morning and evening yields from each and every cow at Burton and as such is regarded as England's first farmers to practice Milk Recording.

Each of the cows at Burton had her own stall for such times as her production was regarded by Mr Evens as adequate. Those animals that failed to meet expectations were quickly identified and having lost their place in the byre, were fattened off for slaughter.

Although not in the position of today's dairy farmers in being able to identify particular traits in potential breeding animals, Mr Evens understood that, generally, 'Like produces Like' and to this end selected stock bulls from deep milkers. The aim at Burton was to develop a herd of Lincolnshire Red Shorthorn cows that combined high milk production with size, quality and constitution.

According to *Stephen's Book of the Farm*, the cows at Burton were fed in such a way as to optimise their genetic potential. In addition to having ready access to fresh water when in their stalls during winter, the cows were given a daily allowance of 4 lb cotton-cake, 2 lb malt coombs, dried grains, 2 lb bran and 3 lb mixed meal (oats or wheat). In autumn these animals received plenty of cabbages and a little later, swedes. During their months indoors, Mr Evens gave his cows two feeds each day with good long hay being provided at night.

When on pasture during the summer months the cows were given 2 lb cotton-cake and if the grass was scarce about 3 or 4 lb of mixed meal would have been given. During this time cabbages or lucerne would have been thrown over the pasture in an attempt to bolster yields. Given Mr Evens' attention to feeding it is perhaps not surprising that during 1905, fifty-four of the Burton cows averaged eight hundred and sixteen gallons and these included several 'one thousand galloners'. By this time, over thirty cows from Mr Evens' herd had won prizes at noted English shows including the London Dairy Show, Royal Counties Show and the Lincolnshire Show.

Mr Evens, who died in 1948, aged 93, also entered cattle at shows on this side of the Irish Sea. At the Royal Dublin Society's annual show in 1905, one of his cows took a first place in the milking trials for the fourth consecutive year. The following year at our Balmoral Show, the aforementioned Miss Katherine Carleton entered a cow in the milking trials that had been bred by Mr Evens. This animal's name was 'Burton Young Cherry'.

Although down through the years a few farmers have dismissed the exhibitions of dairy cows as little more than a 'beauty' contest, the public milking trials offered a cow the chance to demonstrate her 'generosity' at the pail. That is exactly what Miss Carleton's cow did at the 1906 Balmoral show when entered in the milking trial class for 'Cow of any breed or cross exceeding 900 lbs live weight'. When the results of the milking trials were announced the third place went to 'Daisy' owned by George Waring MD of Aghnadarragh, Glenavy, the second place to 'Hawthorne 2nd' owned by Mr James Tutty of 12 Lower Baggot Street, Dublin and the first place (and Challenge Cup) went to 'Burton Young Cherry' owned by Miss Katherine Carleton of Gilford Castle.

Following on from Miss Carleton's decision to retire from dairy farming, a dispersal sale, including several Burton-bred Lincolnshire Red Shorthorns, took place at Gilford Castle on Friday 6 May 1910. The sales notice is reproduced alongside.

Reproduction of notice giving details on Miss Carleton's dispersal sale at Gilford Castle in 1910

IMPORTANT DISPERSION SALE OF HIGH – CLASS DAIRY CATTLE

Messrs John Robson, Limited has received instructions from **Miss CARLETON, GILFORD CASTLE,** who is giving up Dairy Farming to **SELL BY AUCTION** on the **PREMISES** at **GILFORD** on **FRIDAY 6 May 1910,** at **TWELVE** o'clock.

The celebrated Gilford Castle milking herd consisting of **PURE-BRED, SHORTHORNS, LINCOLNSHIRE RED SHORTHORNS, AYRSHIRES, JERSEYS and CROSS-BREDS.** The pure-bred Shorthorns include the famous bull 'Drayton Rearguard', purchased by the Department of Agriculture for 450 guineas. The Lincolnshire Reds, so famous for their milk and butter producing qualities, were nearly all purchased from Mr John Evens, Burton, Lincoln and have been very successful in the show ring and milking competitions. The Ayrshires and Jerseys are also great milkers. Catalogues containing full pedigrees and milk records will be ready in a few days.

Also:

A number of pure-bred Clydesdale Mares; pedigrees in catalogue. Farm implements & etc.

Praising the foresight of Ballycarry's William Calwell

This British Friesian cow was photographed in 1915, the same year that 'Bellahill Moira' was born in Ballycarry, County Antrim.

Just in the same way the village of Magheralin in County Down has its own old song, so too has the County Antrim village of Ballycarry.

"There's a town that is home in a nest among the hills, where the winds drive across sweet and air-y, there she sits like a queen on a throne of emerald green and she's queen of my heart Bal-ly-car-ry".

These are some words from the first verse of the Ballycarry piece, while those of the second verse, acknowledging the lakes and mountains of Kerry, state that Ballycarry was the "Gem of Ireland".

These pastures would have grazed some of Ireland's earliest British Friesians in the 'Bellahill' herd.

Ballycarry village is situated high up on a hill overlooking the patchwork of fields below and also the blue waters of Larne Lough, Belfast Lough and the North Channel beyond. The village, and its surrounding area has a rich history and enthusiasts will be able to describe how the acclaimed author Anthony Trollope (1815–1882) once visited it, and they will take visitors to see Dalway's Bawn, erected around 1609.

The Ballycarry area has a rich farming history as well and *Farm Week* readers may be interested to know that the gentleman who wrote aforementioned song was one of the first agriculturalists to promote British Friesian cattle in the north of Ireland. His name was William Calwell and he was born 30 July 1863.

Around 1881, at the age of eighteen, William Calwell left County Antrim and crossed the Atlantic to work as a builder and architect in California. During his time he was introduced to high yielding American Holstein-Friesian cattle which were descended from Dutch stock. At the turn of the century, one of these deep black and white milkers would have given more than six non-descript Irish cows. According to Mr Calwell at the time a good Friesian would give 2,500 gallons per year against an Irish cow's yield of around 400 gallons.

William Calwell returned to County Antrim in 1910 and over the next few years told local farmers about the great Friesians he had seen during his time in America. One would have thought his message would have been well received but no! When William Calwell described the milking power of the cattle he'd seen in America our farmers proved somewhat dismissive and put it all down to 'Yankee boasting and brag'. In a letter to a provincial newspaper some years later William Calwell recorded that his seemingly "very wild statements about Friesians almost cost me the faith of some of my best friends, who doubted, if not my veracity, then my sanity".

The Department of Agriculture officials took a similar 'stick-in-the-mud' line and when William Calwell tried to import some black and whites from Holland, he found himself 'ham-strung' by governmental restrictions. In the same letter to the newspaper Mr Calwell described these department officials as "obsolete and obstinate old mossbacks!"

Finding his efforts to import stock from Holland hampered, William Calwell persisted and eventually was granted permission to bring some Friesian beasts in from Great Britain. When Mr Samuel Wallace, owner of the 'Knebworth' herd at Swangleys, Herts, held a draft sale on 7 September 1915, Mr Calwell was listed among the buyers.

During the late nineteenth century, when dairy farmers were free to bring in Dutch stock, the Wallace family made full use of the opportunity. By the time that the British Holstein Cattle Society was formed in 1909, one hundred and forty of the two hundred milkers in the Knebworth herd were of Dutch blood. Mr Samuel Wallace was quoted in JK Stanfield's book *British Friesians* as stating that "A good cow costs no more to keep than a bad one and last four times as long". When the Society's first herd book was duly published, ten pages were taken up by the Wallace family's cattle.

One of these animals was 'Knebworth Crissy' (born 1906), the dam of one of the heifers purchased at the aforementioned sale by Mr William Calwell of Ballycarry. This draft sale, which comprised one hundred and sixteen Knebworth females, was conducted by Messrs. John Thornton & Co. These animals average £47 5s 11d and Mr Calwell's 'Knebworth Crissy 2nd' was knocked down for 39 guineas.

At the time of purchase this animal was in-calf and when she gave birth to a heifer calf at Ballycarry three months later, its pedigree details were entered under the name 'Bellahill Moira' in the British Friesian Cattle Society's fifth volume of the herd book. This December born 1915 heifer calf was the forerunner of many fine Friesian cattle bearing the 'Bellahill' prefix. While it was Mr William Calwell who introduced the Friesians to the area, the Bellahill Herd advanced under the direction of his brother, Mr James A Caldwell (different spelling

B. B. FARM, LIMITED,

Ballycarry, Co. Antrim.

———

BELLAHILL MOIRA,

20084

Black and white, white forehead, white over shoulders, white spots on back, white over hind quarters, white switch, calved December 13, 1915, bred by the B. B. Farm, Limited;

Sire, Colton Unity 3717, by Colton Sultan 2525,

Dam, Knebworth Crissy 2nd. 9438, by Knebworth Duke 363,

G.d., Knebworth Crissy 2050.

The pedigree of 'Bellahill Moira' (calved 13 December 1915) as published in the British Friesian Cattle Society's fifth herd book.

of surname) and this latter gentleman's son, Mr David Caldwell

'Bellahill Moira' was not entered in the herd book under the name of William Calwell, but rather that of 'BB Farm, Limited'. This company (the BB stood for Ballycarry/Ballinderry) had been established by Mr Calwell to give his farming friends an opportunity to come together and centrally process their milk into butter and buttermilk by old-time methods. Although the Company faced acute difficulties these were overcome by the good work of its directors, under the inspired leadership of Mr William Calwell. The company became profitable and, in order to become the property of its milk suppliers, went into voluntary liquidation in the early 1920s. Against this background 'BB Farm Ltd', believed to be the only Company in Ireland to manufacture butter and buttermilk by old time method, was transformed into the 'BB Farm Co-operative Society'.

In May 1923 the new company's Presentation committee, which included Thomas Bartley (chairman), John Davision and Jas McDowell (treasurers), and Robert McCully (secretary) recognised William Calwell's tremendous contribution to the Co-operative in particular and the farming community in general. Through a newspaper announcement they recorded how Mr Calwell had "created the dairy, fostered it and watched over it day and night" and how he had "compelled reluctant eyes to see the truth". That month, the committee further marked their appreciation by presenting Mr and Mrs Calwell with a smoker's cabinet and, a solid silver tea and coffee service.

Responding to the presentation through a public notice, Mr Calwell expressed his gratitude for the presentation and recalled how he loved his native village and hoped for a more prosperous Ireland. Stating that the old 'BB Farm Ltd' would not have endured without the self-sacrificing, public spirit of its early members, he added that if the young agriculturalists within the four seas were organised to produce co-operatively, they could sell directly to consumers and have more than 100 million pounds sterling per annum in their pockets. Mr Caldwell also stated if more and more farmers across Ireland went down the road of directly disposing their produce to consumers, "the Ballycarry Co-operators would stand out as the pioneers who blazed the way to Irish agricultural prosperity".

Over the next three decades Mr William Calwell of Ballycarry maintained his deep interest in agricultural matters and took an active interest in the Ballycarry community. He invented the 'Calwell patent hay collector' and when work commenced to build a new primary school at Ballycarry in October 1956, it was Mr Calwell who 'turned the first sod'.

The following year Mr William Calwell of The Bungalow, Ballycarry, passed away on Thursday 30 July 1953, which was the day of his 90th birthday.

Paying tribute to a pioneer of Ireland's Friesian breed

If a hot air balloon had been seen gently floating over the apple orchards of Loughgall in County Armagh around ninety years ago, the basket may have contained a pedigree Friesian bull calf destined to touch down on the pastures of Mr Edward Cowdy's Summerisland Estate.

This 'alleged' occurrence was raised some years ago during an interview with one of the Province's most distinguished British Friesian breeders, whose experience with pedigree cattle spanned seven decades.

Apparently, Mr Cowdy had wanted to bring a Friesian bull calf to his Loughgall Estate but had been told by those in authority that he couldn't bring it in by land or sea. For Mr Cowdy, this was not the end of the matter and the hot air balloon was a way of 'working around' the restrictions in place. Given the controversy surrounding the bull's arrival in Ireland it has, understandably, been rather difficult to uncover hard evidence substantiating the claim.

In addition to being a prominent County Armagh businessman, Mr Cowdy was a leading agriculturalist who played a role in the formation of the UFU. At a meeting held in the Ulster Hall, Belfast on Friday 14 September 1917, he had proposed that the farmers of Ulster should form themselves into a strong organisation to be called the 'Ulster Farmers' Union'. One of the pressing issues at the time involved the low price that farmers were being paid for their milk. Although it was retailing at two shillings a gallon it was the 'middlemen' who, it seemed, were 'creaming off' the profits.

Being a milk producer himself, Mr Cowdy had a vested interest in seeing farmers get a better return for their milk, especially given that he was on the verge of heavily investing in his dairy operation on the Summerisland Estate.

Some months after this aforementioned meeting, Mr Cowdy purchased several expensive British Friesian cows at English on-farm sales. At a time when five good Irish bred Shorthorn-type milk cows would have involved an outlay of £130 at the average Ulster Fair, Mr Cowdy was represented at English sales where he purchased five pedigree British Friesians for 1,630 guineas.

On 18 June 1918, at a sale in Reading, he bought 'Cymric Glossy' and 'Cymric Prudence' for 450 gns and 310 gns, respectively. Eight days later Mr Cowdy paid 200 gns for 'Norton Grand Duchess' at a Norfolk sale and then, during the month of September, he bought 'Thorpe Louise' and 'Thorpe Peerless' for 300 gns and 370 gns respectively, from Mr HW Daking of Thorpe-le-Soken, Essex.

Being a successful businessman, Mr Cowdy wouldn't have speculated in these expensive cattle without hope of getting a good return on his investment. At a time when most of the cows in County Armagh would have been giving around four hundred gallons a year, these black and white cows would yield three times that amount, thereby raising the milk cheque. Undoubtedly when Mr Cowdy bought these cows it would have been in the hope that their progeny would command high prices when the 'Friesian Boom' came to Ireland.

Acting on the old saying that "a good bull is half the herd and a bad one all of it", Mr Cowdy, it would seem, did not undermine his Friesian cows' breeding potential by using a poor bull. Several of the early Friesian calves to be born at Summerisland were sired by 'Routh Bravo' and 'Rough Ringleader Dutchman', two bulls of excellent breeding. Then, in 1922, along came the imported 'Summerisland (imp) Equestrian'.

During the first half of the twentieth century the bringing cattle in from other countries was greatly restricted, especially in the light of recurrent outbreaks of Foot and Mouth overseas. As such, farmers had to make best possible use of those genetics within the national herd. There were occasions when, following negotiations with relevant authorities, cattle could be brought in under strict conditions.

Because of the time lapses between such importations, these animals, when they were finally released, were used to the full. With regard to British Friesian history the influence brought by each of the official breed Society importations in 1914, 1922, 1936 and 1950 has been well documented.

Following on from the successful importation from Holland in 1914, breeders were more than ready to get fresh blood by the 1920s. Finally, after much consultation, a batch of cattle numbering one hundred and twenty-eight was landed in Great Britain. Owing to recurrent disease problems in Holland around that time, these animals (which were of Dutch blood) were selected from South Africa.

When these cattle were sold, they fetched an average of £1,242 each. According to the author JK Stanford OBE, MC, MA (*British Friesians: A History of the Breed*) some of these speculators "later openly regretted their impetuosity". How the aforementioned Mr Edward Cowdy reflected on his purchase in later years is not recorded. At this historic sale the five-month old bull calf 'Summerisland Equestrian' fetched 500 guineas. Having been born on the farm of AA Kingswill, Graaff Reinet, South Africa and brought to Great Britain, this much travelled calf came to Loughgall, County Armagh ... by Hot Air Balloon?

When the Royal Ulster Agricultural Society staged its annual Balmoral Show in 1925 'Summerisland Equestrian' was on parade and listed among the prize-winners. According to a show report the bull had "a lot of quality for his size and great wide quarters".

If any of the farmers attending that particular event had wanted to buy an 'Equestrian'

Old photograph of Mr Cowdy's bull 'Summerisland Equestrian' (imported 1922), born and bred in South Africa.

daughter, they would have been given the opportunity to do so the following month when a major reduction sale took place at Mr Cowdy's Summerisland Estate. It was conducted by the London-based livestock auctioneering institution 'John Thornton and Co' with their Mr Frank Matthews in the makeshift rostrum at Loughgall.

Mr Matthews, who had come to the company in 1887 and 'cut his teeth' selling pedigree Shorthorns, used the sandglass method of closing each deal. When the bidding for a particular beast started to slow down, the sandglass would be turned over. The person whose bidding was topping when the last grains of sand ran through was the successful buyer. Although some people said the sandglass method stopped an unscrupulous auctioneer knocking a beast down early to a friend, the only person who could see the last grains of sand run was the auctioneer!

Before opening the sale, Mr Matthews took the opportunity of addressing those gathered around the ring, many of whom had travelled from distant parts of Ireland. Mr Matthews paid tribute to Edward Cowdy as one of Ireland's pioneers of the Friesian breed and relayed how, seven years earlier, he had been a very liberal buyer at notable English sales. These cattle, stated Mr Matthews, had been purchased with superior judgment at the top of the Boom. Over the years, Summerisland Friesians had won many prizes at provincial shows and, at the time, the annual lactation across all females was 1,080 gallons at 3.7 % butterfat.

Having got the formalities out of the way, Mr Matthews got down to business. Although 'Summerisland Equestrian' failed to meet his reserve of 500 gns, three of the foundation cows, 'Thorpe Louise', 'Thorpe Peerless', 'Cymric Prudence', purchased back in 1918 for 370 gns, 310 gns and 300 gns, fetched 45 gns, 52 gns and 45 gns respectively and although this may have seemed like a rather steep depreciation, Mr Cowdy had taken seven crops of calves off each animal. At the close of business, 35 cows averaged £65 8s and three bulls averaged £45 17s.

Although the prices at Edward Cowdy's Summerisland sale on that day in June 1925 may not have been in the 'Boom' category, the auctioneer Mr Frank Matthews predicted "the future of dairying would be greatly influenced by the Friesian".

The pastures of Summerisland grazed Mr Edward Cowdy's pedigree British Friesians in the 1920s.

THE SUMMERISLAND HERD OF PURE-BRED BRITISH FRIESIAN CATTLE

**JOHN THORNTON & CO WILL SELL BY AUCTION
ON WEDNESDAY 10 JUNE 1925
AT SUMMERISLAND, LOUGHGALL, CO ARMAGH**

ABOUT 50 HEAD OF PURE-BRED BRITISH FRIESIAN CATTLE,
From the well-known Property of
EDWARD COWDY, Esq, DL.
Sale to Commence at 1.30 pm

THE SALE COMPRISES THE ENTIRE HERD, except 10 or 12 head, mostly calves, and includes several heavy milking and noted prize-winning Cows, a number of splendid in-calf and unserved Heifers, and 4 highly-bred Bulls. The sale offers a grand opportunity for Breeders to obtain fashionably bred and typical specimens of this marvellous dairy breed.

Catalogues may be had of Mr E COWDY, Summerisland, Loughgall, Co Armagh; or of JOHN THORNTON & CO, 27 Cavendish Square, London, W1, who will execute commissions.

> "I have seen good cows in all parts of the world, I have seen excellent cows in all parts of the world, but never before have I seen so many great cows together in one group."

Fergus Wilson's Brookmount herd comes under the spotlight

Mr Fergus Wilson's prizewinning British Friesian bull 'Brookmount Captain', born 31 March 1936, out of 'Terling Pheasant 32nd'.

Mr Wilson's Brookmount herd of British Friesian cattle was dispersed at Springfield, Magheragall, Lisburn, on Wednesday 29 March 1950.

SPRINGFIELD, MAGHERAGALL, LISBURN

CATALOGUE

Of Sale by Auction

of

PEDIGREE FRIESIAN HERD

Comprising : MILCH COWS, IN-CALF HEIFERS, MAIDEN HEIFERS, STOCK BULLS, BULL and HEIFER CALVES.

ON THE PREMISES

On Wednesday, 29th March, 1950

Commencing at 10-30 o'clock a.m.

DAVID MAIRS

Auctioneer

8 BRIDGE STREET, LISBURN

Telephone: LISBURN 3138

As a British Friesian cow comes into the make-shift sale ring auctioneer Mr Frank Matthews addresses the crowd, "Ladies and gentlemen, here's another deep milker. Her name's 'Tarvin Indigo' and her details are in your catalogue. Who'll start me off at ninety guineas ... eighty guineas ... seventy guineas surely?" The auction is Mr Edward Cowdy's draft sale of pedigree British Friesians and it is taking place at Summerisland, near Loughgall; the year is 1925.

The bidding for 'Tarvin Indigo' kicks off at forty guineas, and rises in jumps of five guineas to fifty-five guineas and then stops. Mr Matthews picks up his sandglass timing device, "Don't miss her ... the sand is running at fifty-five guineas". Within a few seconds a flurry of bids are coming again before the last grain of sand runs through. Finally, as Mr Matthews brings the gavel down with a sharp crack at sixty-eight guineas, he announces, "Sold to Mr Fergus Wilson of Springfield".

Although this text has been written as a narrative it is all fact based. In 1925 British Friesian pioneer Mr Edward Cowdy did hold a reduction sale at his Summerisland Estate and it was conducted by Mr Frank Matthews of the famous auctioneering House John Thornton & Co of London. It is also true that some years before, Mr Cowdy had purchased 'Tarvin Gladys' for big money in England and on that day in 1925, she was re-sold at Summerisland, by the sandglass method, to Mr Fergus Wilson of Springfield, whose extensive holding ran adjacent to Magheragall Parish Church.

In the case of this particular cow both vendor and vendee had something in common besides an interest in Friesians and that was that they were both prominent businessmen associated with the Linen industry. Mr Edward Cowdy was Managing Director of 'Greenall Mills' and Mr Fergus Wilson was Managing Director of 'Blackstaff Flax Spinners and Weavers', Springfield Road Belfast. Whereas Mr Edward Cowdy's 'Summerisland' herd was the focus of a recent 'Memories from the Farmyard', this week's Mr Fergus Wilson's 'Brookmount' herd comes under the spotlight.

Mr Wilson came to live at Springfield with his sister, Miss Lila Wilson, around 1924. They had a large staff including a parlour maid, housemaid, cook, gardener, handyman and a chauffeur. The role of the latter employee was the subject of some amusement in the local community because Mr Wilson enjoyed driving himself. Often he would have been seen in the car's driving seat 'chauffeuring' his chauffeur in the passenger seat. At Springfield, there were twelve farm workers and before the arrival of tractors, the land was worked by a team of Shire horses.

At the time of purchase the previously mentioned cow 'Tarvin Indigo' was carrying a calf to Mr Edward Cowdy's South African imported bull 'Summerisland (imp 1922) Equestrian'. Subsequently, on 5 November 1925, she gave birth to a heifer calf at Springfield and this was the first animal to be registered in the British Friesian Cattle Society's herd book under the 'Brookmount' prefix. Her name was 'Brookmount Indigo' and having been reared up and put into milk, she remained at Springfield for many years. Visitors to Mr Wilson's farm in

the winter of 1937 would have seen the 'elderly' 'Brookmount Indigo'. That year she'd got off to a great start by giving birth to a heifer calf, on New Year's Day!

Opening the byre door at Springfield, one would have seen a long line of big, deep, black and white Friesian cows. Entering the building accompanied by the Head Cowman, it's easy to imagine him saying, "That one's 'Brookmount Elsie', next to her is 'Brookmount Louise', and then 'Brookmount Gladys'. See that fourth cow, that's a 'Terling' cow". This latter remark would not have been lost on those visitors with even a sketchy knowledge of the breed: What 'Rolls Royce' was to the car industry, 'Terling' was to the dairy industry. Both names were, at a time, synonymous with quality and performance.

The Terling Herd ran alongside that of the equally famous 'Lavenham' prefix and was based in Essex and owned by Lord Rayleigh. Although the extensive Terling Estate was situated in an area of prime arable land, Lord Rayleigh's brother, the Hon Edward Strutt, saw great business potential in building up a large dairy herd supplying milk for the London market. During the latter nineteenth century the cows used were largely crossbred but when Mr Edward Strutt's son, Gerard, came to manage the diary operation, he had the aspiration to develop a herd of pedigree Friesians.

Although his uncle Lord Rayleigh and his father were somewhat sceptical in the initial stages (not to mention askance at some of the prices paid for foundation stock) they were soon 'won over' by the Friesian's milking capacity. In the years that followed, under the direction of Mr Gerard Strutt (1880–1955) and his wife Mrs Rhoda Strutt (1892–1968), the 'Terling' and 'Lavenham' prefixes became famous. Writing about the thousand or so milkers on the Terling Estate, a much travelled foreign breeder once remarked, "I have seen good cows in all parts of

the world, I have seen excellent cows in all parts of the world, but never before have I seen so many great cows together in one group".

'Terling Total Eclipse 30th' was the name of the cow domiciled at Springfield. During the late 1930s, she was a successful member of Mr Fergus Wilson's show team and received high honours at the RUAS Balmoral Spring Show. In addition to being a vice President of the Royal Ulster Agricultural Society, Mr Wilson was a founder member of the Northern Ireland British Friesian Breeders Club, which was formed following an inaugural meeting held during March 1944.

This took place in the Ulster Farmers' Union headquarters of Ocean Buildings, which was situated within a 'stone's throw' from Belfast's City Hall. Sixteen people were present at that historic meeting. A young Sam Wilson of the 'Ravenhill' Herd undertook the duties of Secretary while the 'Brookmount' herd owner, Mr Fergus Wilson, took on the duty of Chairman.

One of the listed aims of the new body was to promote the interests of the British Friesian Breed of cattle and to encourage the keeping of purebred Friesian Cattle throughout Ireland. This would have been something close to Mr Fergus Wilson's heart as he had been one of the early black and white pioneers who struggled to popularise the breed in the face of much opposition.

During a career of breeding pedigree Friesians that spanned twenty-five years, Mr Fergus Wilson registered two hundred and four animals in the Society's herd book. The last 'Brookmount' registration was a bull calf born in the spring of 1950. That year Mr Fergus Wilson left his Springfield residence to reside at Norton, 155 Malone Road. He died on Wednesday 18 January 1957, aged eighty-four years.

The impact of the 'Bursby era' on the British Friesian breed

W hen a person is delivering a speech at a function it is always much more pleasant if the audience like what they are hearing. Take, for example, a talk that Mr WH Bursby gave in the Conway Hotel, Dunmurry, on the evening of Friday 8 November 1963, in which he stated the opinion that pedigree British Friesian cattle breeders were not getting a fair deal.

This would have gone down well with his audience as Mr Bursby was addressing some two hundred and fifty British Friesian breeders. It was the annual dinner of the Northern Ireland British Friesian Breeders' Club, affiliated to the British Friesian Cattle Society, of which Mr Bursby was Secretary.

The reason Mr Bursby felt that breeders were getting something of a raw deal concerned the increasing use of artificial insemination since its introduction around 1946. He felt that the AI centres were getting a lot of money from some widely-used bulls; however, if it wasn't for the pedigree breeders, he argued, the AI centres would not have these great animals. When, at the Conway Hotel dinner in 1963, Mr Bursby suggested that a ten percent surcharge on each AI fee be given back to the breeders, his words would have been received with warm approbation.

'Ballymaguire Bessie 11th' Ex. Born 13 September 1964, sired by 'Salwick Bahrain' and out of 'Ballymaguire Bessie 2nd'.

Mr WH Bursby (1913–1977) had come from a spell in the army to his post as Secretary of the British Friesian Cattle Club in 1946 at the age of thirty-three. Taking up office he was met with a huge backlog of work and set to the task of getting it cleared with military efficiency. These

'Ballymaguire Rebecca 4th', VG. Born 25 October 1964, sired by 'Salwick Bahrain' and out of 'Ballymaguire Rebecca 2nd'.

Ballymaguire Bessie 30th. 2992266, 9/4/69; Salwick Damel 310083 - Ballymaguire Bessie 15th.
Ballymaguire Bessie 32nd. 2999168, 11/5/69; Salwick Damel 310083 - Ballymaguire Bessie 1149512.
Ballymaguire Bessie 33rd. 2999170, 9/5/69; Salwick Damel 310083 - Ballymaguire Bessie 14th. 2420242.
Ballymaguire Bessie 34th. 3007964, 12/6/69; Salwick Damel 310083 - Ballymaguire Bessie 11th. 2262206,
Ballymaguire Bessie 35th. 3065192, 20/9/69; Salwick Bahrain 283623 - Ballymaguire Bessie 18th. 2604092.
Ballymaguire Bessie 36th. 3065194, 3/10/69; Salwick Bahrain 283623 - Ballymaguire Bessie 8th. 1962872.
Ballymaguire Judy 13th. 2990868. 20/4/69; Salwick

Pedigree details on several daughters of 'Salwick Damel' as they appeared in the British Friesian Cattle Society's year book for 1969.

were exciting and turbulent times for the Society. During Mr Bursby's tenure membership increased from less than eight thousand to almost thirteen thousand and the RMX and type classification schemes were introduced. Mr Bursby also took charge when the breed headquarters moved from Central London to Scotsbridge House, Rickmansworth, Herts, during 1953.

When the BFCS celebrated its golden anniversary in 1959 with a succession of special events culminating with a dinner in the Savoy Hotel, Mr Bursby's planning skills were put to the test. When reflecting on these years of intense progress and change, British Friesian breed historians refer to them as the 'Bursby era'.

At the Conway Hotel dinner in November 1963, Mr Bursby was accompanied by the British Friesian Cattle Society's President, Mr Frank Loftus (c.1910–1979) and owner of the famous 'Salwick' herd based in Finchingfield, Essex. Described in a breed journal as being a quiet but persuasive gentleman, Mr Loftus, whose interest in the Friesian went back to

his teenage years, gave valuable support and guidance to the Society over several years.

As a pedigree breeder he was hugely successful, having gone out on his own in 1936 with seventy in-calf heifers and a 'Terling' bull. In the years that followed 'Salwick' blood was used in many herds, indeed the Female Champion at our Balmoral Show of 1947, 1949 and 1950 was by a bull called 'Salwick Baronet', which had come to a county Antrim herd around 1940. When Mr Loftus died in 1979, his obituary printed in the British Friesian Journal stated that "his greatest contribution to the Friesian breed was that he had bred Salwick Dewman."

This bull, born on 31 September 1947, sired daughters with wonderful type and outstanding production figures and, it was generally accepted that 'Salwick Dewman' (died in 1965 at around eighteen years-of-age) was one of the world's greatest sires. His influence was widespread and one of the successful Ulster herds that line-bred to 'Dewman' was the County Tyrone 'Ballymaguire' prefix, run by Mr Thomas Hammond and his son, George, in Stewartstown. Before expanding on this aspect of the Ballymaguire breeding policy, it may be interesting to give some background information on the herd.

Mr Thomas Stewart Hammond (1900–1983) joined the British Friesian Cattle Society in 1955, registering his first heifer in the following year's British Friesian Cattle Society's herd book under the name of 'Ballymaguire Poppy'. She was born 6 December 1955, sired by the Department's AI bull 'Lavenham Trainer' and out of 'Clonmount Poppy 2nd', purchased from the well-known breeder Mr Robert B Martin of Springfield House, Clontonacally, Carryduff.

The Hammond family also owned an older full sister to 'Clonmount Poppy 2nd' called 'Clonmount Poppy', which they took to County

Armagh Agricultural Society's Portadown show in 1959. Although this show was discontinued, in the fifties it drew great crowds. At the 1959 show there were 1,949 entries with classes for forty different dog breeds, Horses, Cattle, Sheep, Swine, Bacon Carcases, Poultry and Home Industries. Children visiting that particular show would have been entertained by the 'Wonderful Wild West Rodeo' and a reconstruction of the popular 'Wells Fargo' Show put on by members of Newry Harriers Hunt Club.

For Mr Hammond and his son George, this 1959 Portadown show was particularly successful as their rising seven-year-old cow 'Clonmount Poppy' 804300 was placed at the top of her class beating JG Hewitt's (Richhill, Co Armagh) 'Ballynahinch Lena' and JF McCall's (Mayfield House, Magheralave Road, Lisburn) 'Magheralave Futurist Ernie' into second and third places, respectively.

In addition to 'Clonmount Poppy' and 'Clonmount Poppy 2nd', several other cows bred by Mr RB Martin's Carryduff herd came to work on the Hammond family's farm in these early years for the 'Ballymaguire' herd. These included 'Clonmount Judy 3rd', 'Clonmount Maureen 2nd' and one called 'Clonmount Rebecca RM' that in her first lactation gave 11,109 lbs at 4.46 in 315 days.

Other bought-in females in the late fifties and early sixties included 'Ravenhill Reality's Amethyst' (bred by Sam and Florrie Wilson, Potterswalls, Co Antrim), 'Rushvale Princess Anne' (bred by Mr SA Caldwell, Rushvale, Ballyclare), 'Forthill Princess Florence' (bred by Mr G Marsden, Fort Hill Farm, Ballycarry), 'Purdysburn Pula' (bred by Belfast Mental Hospital, Saintfield Road, Belfast), 'Oldpark Princess Colleen' and 'Oldpark Princess Lorna' (bred by Miss Winnie Price, Oldpark Road, Belfast) and 'Throne Rosetta 2nd' (bred by

Robert A Ramsey, The Whins, Ballyvesey, Carnmoney).

During these early years of the 'Ballymaguire' herd, two sons and a grandson of the aforementioned 'Salwick Dewman' were brought across the Irish Sea to work as stock bulls. Their names were 'Salwick Nimrod' (born 8 May 1959 out of 'Salwick Nightmare 8th'), 'Salwick Bahrain' (born 18 October 1962 out of 'Salwick Bittersweet 9th' RM, RMX) and 'Salwick Damel' (born 12 September 1965 out of 'Salwick Dorothy 8th' RM, RMX), respectively.

Before visiting Loughry College to inspect a herd of Dairy Shorthorns, members of the Northern Ireland Cattle Breeders' Association visited the Hammond family farm at Ballymaguire during May 1964. The herd, at the time, comprised one hundred and fifty head with around sixty-five animals in-milk. One of the foundation 'Clonmount' cows that had been bought as a calf came in for favourable comment. She had given over two thousand gallons on each of three lactations with one peaking at 2,530 gallons. Speaking at this event Mr Sam Wilson said that Mr Hammond had been "very selective and had aimed to breed animals with size, scope, good wearing ability and good udders".

There was also, at the time of the visit, a beast of Mr Wilson's own breeding that had been purchased by Mr Hammond at the 1957 Allam's Autumn Show and Sale for 120 guineas. This animal's name was 'Ravenhill Lieta'. There were also two cows from Miss Price's 'Oldpark' herd and one from the 'Purdysburn' herd, owned by Belfast Mental Hospital, which was dispersed in 1971.

Most of the 1964-born calves on the ground in the herd at the time of the NICBA visit were sired by either 'Salwick Nimbel' or 'Salwick Bahrain'. Of the two bulls, the latter proved best with many of his heifers growing into great

cows. Two such animals were 'Ballymaguire Ruby 6' and 'Ballymaguire Ruby 7'. They earned a place in British Friesian history becoming the first twin heifers to be classified 'Excellent'. Another daughter of 'Salwick Bahrain' was awarded the Championship at the Royal Dublin Society's Spring Show.

'Salwick Bahrain' himself had an exceptional show career during those years when classes for dairy bulls were still providing a great spectacle at agricultural shows. He was lead out by George Hammond eleven times to take eleven championships. The high point of these was probably the Royal Ulster Agricultural Society's Balmoral Show in 1967, having been put up by judge Mr JM Collie with the Purdysburn Mental Hospital's 'Parkhouse Anther' taking the Reserve Male Championship. 'Salwick Bahrain's other conquests took place at shows held in Ballymena, Antrim, Omagh and Portadown.

Following on the work of 'Salwick Bahrain' in the 'Ballymaguire' herd was the third bull 'Salwick Damel'. Although not taken round the Province's shows, he did in fact become 'something of a celebrity' when Ulster Television was casting for a short Comedy/Drama film called *Boatman Do Not Tarry*. The film, which had been written by John Stewart, was shot in the Bannfoot area of County Armagh. In one scene, actor John McBride had to take a bull across in the ferry and this was where Mr Hammond's stock bull 'Salwick Damel' took up his role in August 1967. The scene was shot and it was subsequently reported that the 10 cwt, one-and-a-half-year-old bull had "given no trouble".

In addition to playing albeit a minor role in the UTV film, 'Salwick Damel' did what a dairy bull is meant to do, sire daughters. The pedigrees of several 1969 'Damel' daughters, as printed in the British Friesian Cattle Society's 59th herd book, are shown in the photograph opposite.

"Although the guest list may have read like a 1930s 'Who's Who' of Ulster agriculture, not all those present were farmers."

Farmers gather in Grand Hotel for dairy awards

Jellied Chicken, Spiced Codfish, Toad in the Hole, Steak and Kidney Pie; these are just some of the delicious main dishes that would have been on the menu of Belfast's Grand Central Hotel during the 1930s. It was regarded by many as one of the city's best establishments and those men and women sampling its hospitality during the 'roaring thirties' could have had a table d'hôte luncheon for three shillings or a table d'hôte dinner for five shillings.

Enjoying their food in a dining room illuminated by electricity, patrons would have appreciated how the ambiance of the hotel was enhanced by some gentle orchestral music. Even those desiring a late night snack were catered for as the Grill Room stayed open until 11.45 pm. Some stewed tripe and onion perhaps … or perhaps not!

On the evening of 13 March 1931, a large contingent of farmers attended the Grand Central Hotel and perhaps, after washing down gingerbread pudding, rich plum cake or some 'dainties' with beakers of fruit cordial, it may have been time to get down to business. This was the annual prize giving of the Central Milk Recording Society of Northern Ireland. Although the guest list may have read like a 1930s 'Who's Who' of Ulster agriculture, not all those present were farmers.

Included in this eclectic gathering were several leading city businessmen and the vice chancellor of Queen's University, Sir Richard Livingstone. Alluding to the presence of these non-farming friends, Captain the Right Hon Herbert Dixon MP stated that their attendance reflected the interest that was being taken in Agriculture throughout the six counties. Although many other industries had, according to Captain Dixon, suffered a 'lean' period during the previous seven years, in the case of agriculture this had been made more acute by a series of wet seasons.

The heavy rainfall did not, however, dampen the enthusiasm of those farmers gathered in the Grand Hotel when it came to distributing the prizes awarded by the Central Milk Recording Society of Northern Ireland. All six counties were well represented and the various cups for 'Best Herd in the County' were handed out as follows.

Armagh – Jas O'Neill, Donacloney Milk Recording Association (seventeen cows averaging 226.36 points), Fermanagh – Mr Wm Johnston, Aghadrumsee Association (five cows averaging 249.87 points), Down – Mr S Patton, Mid Ards Association (nine cows averaging 256.12 points), Derry – Mr MM Mark, Limavady Association (twelve cows averaging 231.01 points), Tyrone – Miss Roche, Doons (thirteen cows averaging 237.35 points) and Antrim – Mr A Ramsey, Ballyclare Association (22 cows 298.16 points).

Having been awarded the most marks, it was the latter farmer, Mr Andrew Ramsey, proprietor of the Throne Dairy based near Belfast's Whitewell Road, who went on to be awarded the Silcock Gold Cup, given to the owner of the 'Best Dairy Herd in Ulster'. Extra points were awarded for 'heifer' lactations in the competition, and since Mr Ramsey's herd included ten such novices, his overall score was given something of a boost.

It was with four of these young females that Mr Ramsey also lifted the Ministry of Agriculture (Northern Ireland) Perpetual Challenge Cup and Gold Medal. This trophy was put up with the objective of encouraging bull owners to retain their animals, until such time as their first crop of daughters had proved themselves in the byre. In the case of Mr Ramsey's four entries, they had given a respectable 9084 lbs (4120 kgs) at 3.9 % butterfat. In his report on that evening's prize

The breeding of pedigree Friesian cattle goes back a long way in Northern Ireland.

d., Findlay Lala 4th. 61470.
28549 Findlay Pat, April 23, 1925; *s.*, Golf (imported) Botermijn 3919, *d.*, Findlay Patricia 38960.
28551 Findlay Patsy, September 14, 1925; *s.*, Golf (imported) Botermijn 3919, *d.*, Findlay Adema Bloom 71186.

The pedigree details on Mr Ramsey's bull 'Old Peter' as published in the British Friesian Cattle Society's Herd Book, Volume 15.

giving in the Grand Central Hotel, the Editor of *Central Milk-Recording Society Year Book* (1931), Mr J Gregg, stated that Mr Ramsey's heifers were all by his "old bull, Peter".

Had Mr Ramsey wanted to be pedantic, he could have pointed out to Mr Gregg that 'old Peter' was in fact, a highly bred and five times Balmoral Show prize-winning bull, correctly known as 'Findlay Peter' 28553. Furthermore, Mr Ramsey could have stated that the bull's breeder, Mr George Findlay of Stonehaven, Kincardineshire, had entered the animal's pedigree details in the British Friesian Cattle Society's fifteenth herd book. Born on 10 March 1925, 'Findlay Peter' sired by 'Findlay Footprint' 19999 and out of a cow called 'Findlay Polly 3rd', had an interesting pedigree, his paternal grandparents being 'Golf (imp) Botermijn' and 'Findlay Clara 2nd'.

Old Peter's or rather 'Findlay Peter's grandsire 'Golf (imp) Botermijn' had come to Great Britain as a yearling calf as part of the British Friesian official importation from Holland in 1914. He had been bred by GR Miedema of Leeuwarden and was purchased by Tadcaster breeder Mr John Bromet for £525. In 1917 'Golf (imp) Botermijn' changed hands for 1,700 guineas and then again in 1919 for 2,700 guineas!

'Findlay Clara 2nd', maternal grand-dam of Mr Ramsey's bull, was something special too and was recorded as having given thirteen gallons in one day. During 1925 this fine cow gave an incredible 3616 gallons, but as sometimes occurs with valuable animals, she was cut off in her prime having taken lead poisoning after the calving.

Mr Ramsey would have been aware of the impressive linage of this bull at the time of purchase. 'Findlay Peter' combined good breeding with good looks, having been paraded at the Balmoral Spring Show on five consecutive occasions. His debut on the Society's cattle lawn took place in 1927, when he was awarded a second place to a bull that has previously been profiled in *Farm Week*'s 'Memories from the Farmyard'. This bull was called 'Parks Marthus 3rd', exhibited by Purdysburn Mental Hospital, Belfast.

The following year, at three years-of-age and 23¾ cwt in weight, 'Findlay Peter' was back at Balmoral to take the Supreme Championship in the Black and White section. Having been fed almost solely on Clarendo, the feeding stuff's manufactures were keen to use a quotation from Mr Ramsey in their marketing campaign. They couldn't have sought endorsement from a better man and were able to quote him as having said, "People using other feed stuff were only losing their horses' and mens' time in carting it".

Having been placed at the top of his classes in the following shows in 1929 and 1930, 'Findlay Peter's winning run at Balmoral came to an end in 1931 when he was, one could say, back were he started with a second prize ticket. He was, however, beaten by a worthy champion, turned out by a man destined to become President of the Ulster Farmers Union between 1944–45 and 1949–50. His name was Mr AE Swain, CBE, New Orchard, Moira, and his bull was 'Summerisland Koko' (born 21 February 1929 by 'Summerisland (imported 1922) Equestrian' and out of 'Summerisland Lorna').

Those farmers visiting Mr Andrew's Ramsey's herd late in the year of 1931 may have been interested to see 'Findlay Peter' on the Throne

Name of Calf	Date of Birth	Name of Dam
Throne Darling	18 December 1927	Throne Megpie 3rd
Throne Peggy	20 December 1927	Throne Dora 2nd
Throne Kate 4th	4 November 1928	Throne Kate
Throne Meg 4th	10 February 1928	Throne Meg
Throne Freda	5 December 1928	Throne Meg 2nd
Throne Merrylass	2 November 1928	Throne Megpie 4th
Throne Esther	12 April 1929	Throne Megpie 2nd
Throne Jane	28 April 1929	Throne Winnie
Throne Gift	15 April 1930	Throne Queenie
Throne Lizzie	22 January 1930	Throne Belle
Throne Martha	5 April 1930 (Twin)	Throne Winnie
Throne Prospect	5 April 1930 (Twin)	Throne Winnie
Throne Winnie 6th	17 June 1931	Throne Winnie
Throne Freda	19 June 1931	Throne Freda

Daughters of Findlay Peter born in Mr Andrew Ramsey's Throne herd

farm. Undoubtedly, they would have had a greater preoccupation with his daughters in the knowledge that a dairy bull is only as good as 'what he puts on the ground'. The visitors would not have been disappointed. Pointing out some of Old Peter's first calvers, Mr Ramsay may well have said, "Back in the spring, I was awarded a cup in the Grand Central Hotel for those ones."

During the 1940s and 1950s Mr Andrew Ramsey continued to register both Ayrshire and British Friesian cattle in their respective herd books. After a long and influential career as a cattle breeder, Mr Andrew Ramsey of the Throne Dairy, Whitewell Road, Belfast, passed away on 16 September 1948.

"It's good to give a young heifer the experience of a show, in preparation for that time when she's in her prime."

Depth of breeding bears fruit for Ravenhill herd back in mid-60s

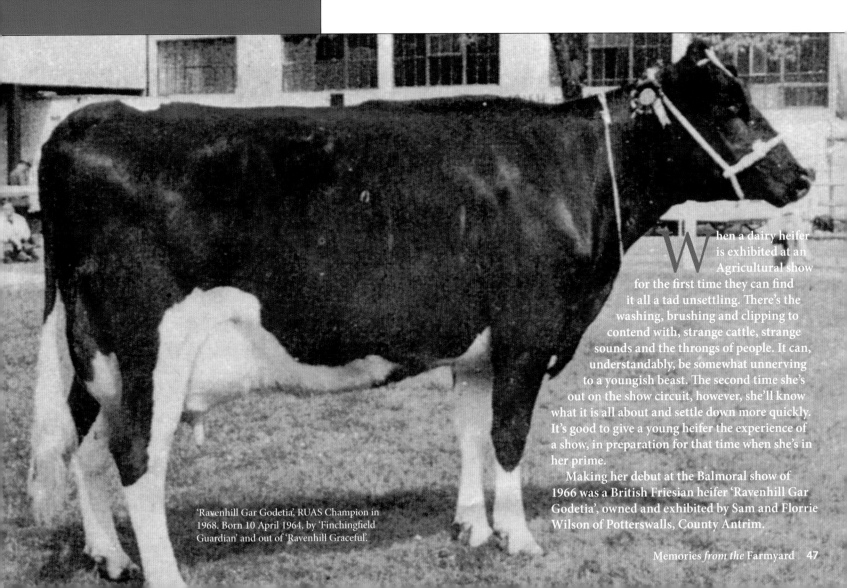

'Ravenhill Gar Godetia', RUAS Champion in 1968. Born 10 April 1964, by 'Finchingfield Guardian' and out of 'Ravenhill Graceful'.

When a dairy heifer is exhibited at an Agricultural show for the first time they can find it all a tad unsettling. There's the washing, brushing and clipping to contend with, strange cattle, strange sounds and the throngs of people. It can, understandably, be somewhat unnerving to a youngish beast. The second time she's out on the show circuit, however, she'll know what it is all about and settle down more quickly. It's good to give a young heifer the experience of a show, in preparation for that time when she's in her prime.

Making her debut at the Balmoral show of 1966 was a British Friesian heifer 'Ravenhill Gar Godetia', owned and exhibited by Sam and Florrie Wilson of Potterswalls, County Antrim.

THE DAM LINE OF RAVENHILL GAR GODETIA, BORN 10 APRIL 1964

Dam	Ravenhill Graceful	Born, 1 September 1954, bred by the Wilson family, County Antrim.
G dam	Ravenhill Reality's Greta	Born 17 March 1952, bred by the Wilson family, County Antrim.
3rd dam	Ravenhill Margretta	Born 7 March 1947, bred by the Wilson Family, County Antrim.
4th dam	Fairyknowe Martha	Born 12 July 1935, bred by Mr McVeigh, Whitewell Road, Belfast.
5th dam	Throne Daffodil	Born 5 January 1933, bred by Mr Andrew Ramsey, Throne Dairy, Whitewell, Belfast.
6th dam	Throne Flossie	Born 21 March 1928, bred by Mr Andrew Ramsey, Throne Dairy, Whitewell, Belfast.
7th dam	Ballyhill Savoureen 2nd	Born 2 January 1923, bred by James A Caldwell, Bellahill House, Carrickfergus.
8th dam	Bellahill Shamrock 3rd	Born 22 January 1919, bred by James A Caldwell, Bellahill House, Carrickfergus.
9th dam	Bellahill Moira	Born 13 December 1915, bred by BB Farm, Ballycarry, County Antrim.
10th dam	Knebworth Crissy 2nd	Born 28 November 1911, bred by Samuel Wallace, Knebworth, Herts.
11th dam	Knebworth Crissy	Born in 1906, bred by Samuel Wallace, Knebworth, Herts.

Full of promise, this young animal came from a family of Balmoral winners, indeed her grandmother had been the female Champion in 1960 and a full sister to her great, grandmother, female Champion at the Balmoral shows of 1947, 1949 and 1950.

Given this background, Sam and Florrie Wilson would have had much hope for their 'Ravenhill Gar Godetia', entered in the class for 'Heifer, born between September 1963 and August 1964' at the Royal Ulster Agricultural Society's Spring Show of 1966. Before looking at how the heifer did at this, her first 'Balmoral', it may be interesting to give some more information regarding her background.

It was during the month of August 1961 that *Farm Week* journalist Mr EJ Sloan paid a visit to the Wilson family's farm at Potterswalls on the outskirts of Antrim. This was home to the Ravenhill herd of pedigree British Friesians (est. 1935) that, at the time of Mr Sloan's visit, comprised 115 animals including around forty milkers.

The 'latest acquisition' at the time of Mr Sloan's visit was a yearling bull, jointly purchased with Messrs Magee Bros, Newry. It was a son of the famous 'Salwick Dewman' (b.31 December 1947), an English bull that during a lifetime of around eighteen years, sired one hundred and sixty-three RM daughters, sixty-nine RMX, eighteen classified 'Excellent' and three consecutive female champions at Royal Agricultural Society of England shows.

The young 'Dewman' son jointly purchased by the Wilson family and Magee brothers, was called 'Finchingfield Guardian', born on 4 June 1960 out of a cow called 'Finchingfield Collona 5th'. Between 1961 and 1966 'Finchingfield Guardian' was used in the Wilson's herd both naturally and by Artificial Insemination. Those of his daughters to be classified 'Excellent' included 'Ravenhill Gar Bud', 'Ravenhill Gar Adele', 'Ravenhill Gar Amphoria' and 'Ravenhill Gar Oread'.

When the aforementioned heifer, 'Ravenhill Gar Godetia', made her debut at the 1966

Balmoral she was twenty-three months of age and a day or two off the calving. Having spent the first night in Balmoral's Londonderry Hall, she was brought to the parade ground's Ring 5 on Wednesday 26 May 1966 where Mr EE Davies of Cliftonmill Farm, Rugby was judging. Having made his deliberations, Mr Davis placed 'Ravenhill Gar Godetia' at the top of her class, with the second and third tickets going to two 'Barbican' beasts called 'Barbican Dew Rosaleen 9th' and 'Barbican Dew Judy 7th'. It was a great start to the Ravenhill heifer's show career … and there was more to come.

When the Royal Ulster Agricultural Society staged its 101st Balmoral show on 22, 23, 24 and 25 May 1968, the Wilson family of Potterwalls turned out a strong team in the British Friesian classes. They took three first, two second, and two third prizes and their team included the 'remarkably spry', thirteen-year-old 'Ravenhill Lena' (calved 14 June 1955, by 'Ravenhill White Heir' and out of 'Ravenhill Princess Mina'), which came second in a very strong class of seventeen dry cows. The 'star' of the show, however, was the aforementioned 'Ravenhill Gar Godetia' that had taken the heifer class in '66.

By this time, 'Ravenhill Gar Godetia' was rising four years-of-age and accustomed to the environment of an agricultural show. On the morning of the parade she came out looking her best and was eventually awarded the top honours with Mr G Marsden's 'Forthill Jans Jollona' (calved 3 September 1962, by 'Horwood Collona Jan' and out of 'Forthill June', RM, RML) taking the Reserve.

Describing his selection as champion at the 1968 Balmoral the judge, Mr John Craig, stated that 'Ravenhill Gar Godetia' was "an excellent animal and an outstanding example of good breeding", adding that "she had very few faults and could be compared favourably with anything across the water".

The birth of classification and the first 'Excellent' cow in Ireland

Born in 1951, 'Ravenhill Princess Gentian' (in the foreground) was the first cow in Ireland to be classified 'Excellent'.

When the council of a society or association is trying to sell a new concept to grass roots members it is always useful if they, shall we say, 'wheel out the big guns'. This was the case when leaders within the British Friesian Breeders' Club organised a meeting in November 1959, with the intention of gaining support for the introduction of a 'Type Classification Scheme'. Two of the platform speakers at that historic gathering were Mr Robert Rumler, Executive Secretary of the Holstein Friesian Association of America and Mr George Clemons, Secretary Manager of the Holstein Friesian Association of Canada.

After winning the Reserve Female Championship at Balmoral show in 1965, 'Ravenhill Princess Gentian' gave birth to a bull calf in the Londonderry Hall. It was named 'Ravenhill Genitor'.

Exactly one year later 'Ravenhill Genitor' was back at Balmoral where he got a first place in the class for 'British Friesian Bull, calved on or after 1 September 1964'.

'Ravenhill Genitor' was used as a stock bull in the Wilson family's 'Ravenhill' herd until August 1969. Several of his daughters won championships at various show and sales. These included 'Ravenhill Gentor Naza' (Banbridge 1970), 'Ravenhill Gentor Mina' (Banbridge 1971) and 'Ravenhill Gentor Phora' (Banbridge 1972).

NAME OF CALF	SEX	DOB	SIRE
Ravenhill Gesture	Bull	8/8/1954	Ravenhill Magnus
Ravenhill Generator	Bull	20/8/1955	Ravenhill White Heir
Ravenhill Kinton Esther	Heifer	29/9/1956	Ravenhill Kinton
Ravenhill Genton	Bull	3/11/1957	Ravenhill Kinton
Ravenhill General	Bull	23/11/1958	Standalane Rommel
Ravenhill Herta Genista	Heifer	3/1/1960	Ravenhill Greatheart
Ravenhill Herta Gina	Heifer	9/1/1961	Ravenhill Greatheart
Ravenhill Herta Geum	Heifer	8/3/1963	Ravenhill Greatheart
Ravenhill Generous Ravenhill Genial	Twin Bulls	16/3/1964	Finchingfield Guardian
Ravenhill Genitor	Bull	27/5/1965	Finchingfield Guardian

The meeting, which was attended by a representative of the Northern Ireland British Friesian Breeders' Club (formed 1944), took place in the Corn Exchange Hall, Reading. Having spoken about the experience of North American breeders, Mr Rumler exhorted those gathered to embrace a scheme by which the breed's cows and heifers could be classified. "Without hesitation or reservation", he stated, "it is the thing you should be doing and I hope it brings as good results as it has done in North America".

Outlining the Canadian Classification scheme which dated back to the 1920s, Mr Clemons said that before its existence, dairy cows were evaluated almost entirely on their capacity to give a big volume of rich milk over a short time. This did not give a true reflection of the animal's worth, as often because of body defects such high yielders had to be culled at an early age. There were cows with deep milking powers that had to be sent for slaughter due to bad feet and legs or too pendulous an udder.

Addressing this problem in Canada, work was undertaken to promote dairy stock with the conformation (type) that would withstand the rigours of production over a long period. When a consensus was achieved as to how a good Holstein should look, the services of a sculptor were employed to make up a 'true-type' model. During the 1950s Canadian Holstein Friesian breeders were using score cards and this represented a major step forward in forming the basis for a classification system. When drawn up and agreed, the groups were as follows, 'Excellent' 90–100 points, 'Very Good' 85–90 points, 'Good' 75–84 points, 'Fair' 65–74 points and 'Poor' below 65 points.

Back in Great Britain and Ireland there was, towards the end of the 1950s, a growing interest in type classification within the ranks of the British Friesian Breeders' Society (est. 1909), to which the Northern Ireland Club was affiliated. One of the nation's most progressive breeders at that time was Mr Ben Cooper of the famous 'Normead' prefix based in Winterbourne Monkton, Swindon. During the late 1950s, he visited Canada to study the Classification method. On returning home Mr Cooper endorsed the scheme by speaking at various breeders' meetings and writing articles for the bi-monthly British Friesian Journal.

It was seen by many as a major step forward when the British Friesian Society's council took the decision to introduce its own classification scheme, which would come into effect on 1 March 1960. It was to be similar to those operating across the Atlantic but with some modifications to suit the needs of British and Irish pedigree breeders. There would be six classes, 'Excellent' 90–100 points, 'Very Good' 80–89, 'Good' 70–79, 'Average' 60–69, 'Fair' 50–59 and 'Poor' under 50 points.

Even before the task of classifying the cows on farms began, over 5,000 applications were received. One of the Society's classifiers, Mr Austin Jenkins, used the British Friesian Journal of June 1960, to enlist co-operation from participating farmers. He stated that with an efficient approach, between forty and fifty beasts could be classified in a day and under ideal conditions the animals would be inspected in an enclosed concrete yard adjacent to the cow-house. Mr Jenkins also pointed out that the field officers would be using rubber ink stamps on documents and as such a table would be needed. If a table wasn't available he added, "an old door on some straw bales" would suffice.

On 2 September 1960 a bought-in twelve-year-old cow called 'Lavenham Welcome 166th' RM was inspected on the farm of Mr RE McKendrick of Cameron Farm, Windygate, Fife. Although farmers had been warned about having too high expectations for their 'pet' cows, Mr McKendrick must have had great hope awaiting the classifier's result when it came to 'Welcome 166th'. She had been a female Champion at top shows including 'The Royal' and 'Royal Highland'. Mr McKendrick's aspirations would have been well placed, and it was a great cause for celebration when it was announced that that 'Lavenham Welcome 166th' was to be the first 'Excellent' cow in the United Kingdom and Ireland.

Regarding classification in Northern Ireland, early participating British Friesian herd owners included Mr SJ Dickson, Hillside, Derryboy, Crossgar ('Crossgar' herd), Mr Thomas W Hammond, Ballymaguire, Stewartstown ('Ballymaguire' herd) and Samuel Wilson & Co, Potterswalls, Antrim ('Ravenhill' herd).

Several of the first cows to classify in the latter herd were by a Scottish bull called 'Craigiemains Tiptop' that the Wilson family had purchased around 1946. This bull had been sired by 'Craigiemains Prince Albert' and perhaps this is why the pedigree names of 'Tiptop's daughters included the word 'Princess'. Those first cows to be classified on the Wilson's farm at Potterswalls, included 'Ravenhill Princess Blossom', 'Ravenhill Princess Enchantress', 'Ravenhill Princess Irene' and 'Ravenhill Princess Mina'. Following assessment by the British Friesian Society's field officer, each of these cows was awarded the 'Very Good' class.

There was, however, another 'Tiptop' cow classified at that time and her name was 'Ravenhill Princess Gentian', born on 14 December 1951 out of 'Ravenhill Corona' by 'Salwick Baronet'. During a long breeding life in the herd of her birth, the 'Gentian' cow produced ten registered calves (two of which were used in the 'Ravenhill' herd as stock bulls) and gave over fifty tons of milk. Not only was she successfully shown as a three-year-old at the Royal Highland Show in 1954, she also won top honours at Dublin and Balmoral.

All in all a great cow and now *Farm Week* readers may be asking, "How did she classify?" Well, suffice to say that when the British Friesian Society's official took their leave of Potterswalls back in 1962, 'Ravenhill Princess Gentian' had earned a special place in history, becoming the first cow in Ireland to be classified 'Excellent'.

Florrie was at the forefront of development of Friesian breed

Over twenty years have passed since our farming community lost the highly respected Miss Florence Margaret Wilson (1919–1987) who from many years ran the famous 'Ravenhill' herd of pedigree British Friesian cattle with her brother Mr Sam Wilson (1922–1999).

At fourteen years-of-age Florence Wilson, known affectionately as 'Florrie', left school and started milking the Shorthorn type cows belonging to her father Mr Samuel Wilson on their Ravenhill farm at Crumlin, County Antrim.

During the mid 1920s Mr Wilson had the opportunity of using a neighbour's British Friesian bull. Having put some of the resulting half-bred progeny into milk with good results, his interest in and desire to work with Friesian dairy cows was stimulated.

Florrie Wilson, her mother Mrs F Wilson and Sam Wilson junior, at Potterswalls in 1966. The cows had a combined age of 44 years and had yielded well over 150 tons of milk between them. The cow to the right of the picture was the first 'Excellent' cow in Ireland.

In 1933, Samuel Wilson and his wife paid a visit to the farm of Mr Thomas McVeigh, Fairyknowe, Whitewell Road, Belfast, with the intention of buying some Red Island cockerels. When they were there, Mr McVeigh asked if they'd like to buy a pedigree Friesian bull calf. A deal was struck and the animal, named 'Fairyknowe Comrade' (born 10 June 1933) came to the Ravenhill farm.

Two years later in 1935, Samuel Wilson sold five commercial cows to buy three pedigree Friesian heifers from Tom McVeigh. Their names were 'Fairyknowe White Blossom', 'Fairyknowe Flora' and 'Fairyknowe Honey' and they cost £18, £20 and £20, respectively.

On 14 November 1935 the latter beast dropped a heifer calf and this was first animal to be entered in the British Friesian Cattle Society's herd book under the 'Ravenhill' prefix. Over the next 50 years there would be eight hundred and thirty-one more! This newborn heifer's details were published in the Society's 25th herd book under the name 'Ravenhill Sweetie'.

In 1939 the Wilson family bought another animal from Tom McVeigh and her name was 'Fairyknowe Martha'. A great asset to the fledging 'Ravenhill' herd, this cow was the mother of 'Ravenhill Mariette', which was Royal Ulster Agricultural Show Female Champion at Balmoral in 1947, 1949 and 1950.

The 'great' 'Mariette', whose portrait hung in the Wilson family home for many years, was the first of six 'Ravenhill-bred' cows that were Champions at Balmoral. The others were 'Ravenhill Bonnie Breeze' (1953), 'Ravenhill Reality's Greta' (1960), 'Ravenhill Nor Tata' (1965 and 1966, under the ownership of Mr Hugh Magee), 'Ravenhill Gar Godetia' (1968) and 'Ravenhill Gar Petal' (1972).

During an interview with the writer in 1994, Sam Wilson stated that although he was often regarded as the 'public face' of the 'Ravenhill' herd and was most often seen at the shows and, show and sales during the year, this was because his sister Florrie was at home "doing all the work". She was, however, a willing worker and was reported as having said, "I don't know what I'd do with myself if I didn't have cows to milk".

That Florrie Wilson liked cows there can be no doubt but she did, according to Sam, "only like good ones". If there was one that didn't measure up, be it for type or temperament she'd say to Sam, "It's time that one was away ... and quickly!" Having described her perfect beast in a newspaper interview, Florrie Wilson concluded by quoting a line from a Keats poem, *"A thing of beauty is a joy forever"*.

When the Northern Ireland British Friesian Breeders' Club was formed in 1944, Florrie Wilson was one of the founding members and, over the years, worked tirelessly to progress its work in terms of breed promotion. In 1969, when the members celebrated their club's Silver Anniversary to mark the passing of twenty-five years, it was highly appropriate that Florrie Wilson was serving as Club President.

That year's celebrations coincided with the sixty-year commemorations of the British Friesan Breeders Society, which was formed in 1909. In a speech delivered at a special dinner held in Belfast Club, President Florrie Wilson stated, "This is our silver anniversary, the parent body's diamond jubilee and I hope it will be a golden one for the British Friesian breed".

Given that the membership of the Northern Ireland British Friesian Breeders' Club had risen from one hundred and eighty-three in 1948 to six hundred and fifty-two in 1969, this was a great time to reflect on all that had taken place.

The meeting was, however, tinged with sadness concerning Father Patrick Collins, who had been Director of studies at the Warrenstown Agricultural School in County Meath. A great supporter of the Friesian breed, he had been due to take up the post of President of the British Friesian Cattle Society but had passed away one month prior to being installed. Had Father Collins lived long enough, he would have been the first Irishman and the first priest to become Society President.

Addressing those gathered in Belfast, Florrie Wilson stated that Irish breeders and indeed the whole British Friesian Society throughout the British Isles would be very much the poorer from his passing. Within a decade or so of this meeting, Florrie's own brother Sam, would become the first person from Northern Ireland to serve as President of the British Friesian Cattle Society, which at the time had over fourteen thousand members.

In addition to her work with the Friesian Club and Methodist Church during the 1970s, Florrie Wilson became a governor of the Gurteen Agricultural College, based in County Tipperary. She attended meetings with regularity and her input was greatly appreciated. According to fellow council member, Rev Dudley Cooney, Florrie "had great knowledge and wisdom and spoke her mind with candour and kindness".

Everyone who knew Florrie Wilson appreciated her great enthusiasm for life, her sharp wit, strong faith, sound common sense and also the fact that she took little nonsense and could deliver a 'gentle' chiding if necessary. This later aspect was relayed to the writer some years ago when undertaking research on the Ravenhill herd.

One day Florrie was cleaning out the pen of a young bull that was being reared up for private sale as a breeding bull. Although little more that a lump of a calf, the bull started to shake its head and paw the ground. Somewhat taken aback by this display of insolence in such a youth, Florrie cuffed him with the brush and in a loud voice exclaimed, "I'll not be for taking that off the likes of you!"

"Well I suppose that 'Ravenhill Val Verna' was the most internationally known".

Paying homage to the greatest of the famous Ravenhill cows

When the late Sam Wilson of the 'Ravenhill' herd was asked during an interview in 1994 to name his family's most successful British Friesian cow, he hesitated before answering. He could have mentioned one of several animals starting off with the three times Balmoral Champion 'Ravenhill Mariette' that took the top honours in 1947, 1949 and 1950. 'Ravenhill Bonnie Breeze', 'Ravenhill Reality's Greta', 'Ravenhill Gar Godetia' and 'Ravenhill Gar Petal' were all Balmoral winners too, having scooped the breed's female championship in 1953, 1960, 1968 and 1972.

Perhaps, as Sam pondered the question, his 'Ravenhill Princess Gentian' sprang to mind as Ireland's first 'Excellent' classification or maybe 'Ravenhill Princess Enchantress', whose show career continued into her thirteenth year and then there was the predominately white cow 'Ravenhill Lena', a 'star' of Ballymena show. Or the record-breaking 3,200 guineas 'Ravenhill Stardust', another headline maker.

Sam could have nominated any of these cows but he didn't. After giving the matter careful consideration, he tapped the ash off the end of his cigarette and said, "Well I suppose that 'Ravenhill Val Verna' was the most internationally known".

On Monday 26 July 1976, members of the Northern Ireland British Friesian Breeders' Club (est. 1944) and their friends attended a Field Day at Gartree on the shores of Lough Neagh. This was home to Sam and Florrie Wilson's Ravenhill herd which had, some three years previous, moved from Potterswalls, near Antrim.

Among those gathered at Gartree on that day were Mr JK Lynn, Chairman of the Northern Ireland Milk Marketing Board, Mr Harold Adams, Northern Ireland British Friesian Club President and Mr Robin Morrow, President of the Ulster Farmers' Union. Also present was Mr Morrow's son, Peter, who was the winner of a competition in which visitors ranked six cows in order of merit.

One of the cows at Gartree at the time of the 1976 Field Day was the aforementioned 'Ravenhill Val Verna'. Few people, including Sam and Florrie Wilson, could have envisaged what lay ahead for this cow. At the time of the Field Day 'Val Verna' was four years-of-age and six weeks off calving.

Before providing more information on how she went on to gain an 'international reputation', it may be of interest to touch in on her pedigree that went back sixteen generations to a cow called 'Terling Breeze' 3936 and was entered in the Society's first herd book published in 1912. Indeed, eleven of 'Ravenhill Val Verna's ancestors were all 'Terling' cows. Her fifth dam had come to Ravenhill in 1950 and was the last bought-in female to join the herd.

Born on 24 March 1972, 'Ravenhill Val Verna' had been sired by a homebred bull called 'Ravenhill Valorum'. One visitor to Potterswalls recalled visiting the holding just before the move to Gartree, and how Florrie Wilson asked her brother to "Go and get Valorum". Within a few minutes Sam led the stock bull out of his stone-built house and into the yard. He had good size, was predominately black and although not maintained in show condition himself, 'Ravenhill Valorum' did sire several prizewinning daughters including 'Ravenhill Val Nora 2nd', purchased at Banbridge in 1975 by Mr Roger Shipsey of Waterford for 900 guineas.

Having completed his term at Ravenhill, 'Valorum' spent some time working in George Kingston's noted 'Cradenhill' herd based in County Cork were, according to a British Friesian Journal report, he "proved a great bull for tightening udders". Following this, Valorum was on the move again and was taken across the Irish Sea to join the 'Salwick' herd in Essex. Sometime later his daughter, 'Ravenhill Val Verna', also crossed the water, not to be milked

'Ravenhill Val Verna' (born 24 March 1972) out of 'Ravenhill Vera' and sired by 'Ravenhill Valorum' was the winner of the 'Best Exhibitor Bred' award at the fourth World Conference Show held in Stoneleigh, Warwickshire, during October 1976.

on a dairy farm but rather to be exhibited at the fourth World Friesian Conference.

This important event took place every four years. The first had taken place in Amsterdam during 1964, the second in Canada and the USA during 1968 and the third in Italy during 1972. All these events had boosted Black and White breeding across the globe and it was with an air of optimism and a degree of trepidation, that the honour of organising the event came to Great Britain. The first journal of 1976 stated, "This year brings the fourth World Conference and the British Friesian Cattle Society of Great Britain and Ireland is privileged to act as host for this major international event." It was scheduled to open on Monday 11 October 1972 at Stoneleigh.

In an attempt to raise funds for the 1976 World Conference event, members of the Northern Ireland British Friesian Breeders' Club held a special calf sale, following a suggestion by Mr Trevor Gibney of Newry. It was decided that members be invited to sell a calf at Banbridge Livestock Ltd on Wednesday 17 March 1976 and donate half of what was received to the parent Society. Mr George Bryson conducted the sale of calves which sold to a top of £350 paid for a 'youngster' from Mr Gibney's 'Ballynanny' prefix.

The thirty-five calves offered made £6,835 thereby generating £3,417 to be forwarded. This sum was further increased by generous cheques from John Thompson & Sons and the Northern Bank. Additional money was also received from several members who were unable to bring calves. The effort by Northern Ireland British Friesian Club members was much appreciated across the water. The May 1976 Breed Journal stated that it had been "a great day, created by a group of great people with whom the Society was proud to be associated".

Given the Northern Ireland Breeders' contribution, it was most fitting that, having obtained the co-operation of the Northern Ireland Ministry of Agriculture, six of our Friesians were sent to the World Conference Show. The group was made up of four animals from Mr David Heenan's 'Barbican' herd and, two from the Wilson family's 'Ravenhill' herd. As previously stated one of the Northern Ireland exhibits was 'Ravenhill Val Verna'.

At Stoneleigh the cattle were stalled in the Society's marquee and were the focus of much attention. On the third day of the week, 'Ravenhill Val Verna' was led out to compete in a class of forty-five for the 'Cow in-milk, calved three times' class. When the places were made known, the first place went to Mr and Mrs K Stafford-Smith's 'Shopland Edleet Ruth 6' and, to the applause of many, the second place to the Wilson family's 'Ravenhill Val Verna'. As the Shopland cow had been bought-in by her owners, it left the way open for Val Verna to lift the 'Best Exhibitor-bred' entry in the class.

At the time this fourth World Conference event was taking place, Northern Ireland farmers were not permitted to import female cattle and as such, 'Ravenhill Val Verna' had a 'one-way ticket'. After the show, she went to join the famous Somerset based 'Sharcombe herd', owned by Mr Keith Showering.

Following her success at the Conference event, 'Ravenhill Val Verna' joined the 'Sharcombe' show-team and during 1977, 1978 and 1979 won successive prizes at the Royal Show, Royal Bath and West, Royal Welsh, Three Counties, South of England and Royal Bath and West. At the North West Club's jubilee show at Bingley Hall, Stafford, she was re-united with her breeder. When Sam was relaying this 'meeting' during his interview in 1994 he laughingly stated that when he went up to 'Val Verna', she seemed to recognise him!

In this column each week, we always strive to end each and every piece on a positive and optimistic note. It is not, however, always possible. Apparently while still in her prime, the 'internationally known' 'Ravenhill Val Verna' broke out, ate a toxic plant, and died!

Name of calf	Sex	Date of birth	Sire
Ravenhill Ranee	Heifer	7 July 1974	Ravenhill Aurelius
Ravenhill Verena	Heifer	6 June 1975	Weeton Classic 2
Ravenhill Ven Vena	Heifer	14 September 1976	Largyvale Revenue
Sharcombe Valpollcella	Bull	30 August 1977	Finalex Norman
Sharcombe Vina	Bull	11 September 1979	Normead Bullion

Details on registered British Friesian calves out of Ravenhill Val Verna

> *"Back in the 1940s, when less milk recording information was available, farm visits were all the more meaningful."*

A new optimism at Ayrshire sale in the wake of World War II

When it comes to buying-in young stock from other herds many milk producers have, down through the years, found considerable benefit in visiting the home farm of potential recruits. There the cattle could be inspected, with their relatives, in their 'working clothes'. For instance, before buying a nice batch of maiden heifers by a certain bull, it's always good to see a few 'in-milk' beasts by the same sire. Such an opportunity, one could say, broadens the prospector's perspective.

Back in the 1940s, when less milk recording information was available, farm visits were all the more meaningful. Suppose, during this decade, a farmer had been considering buying a few pedigree Ayrshire heifers from Cecil McCormick's fine 'Willow Bank' herd, they may well have travelled to his home off the Knockbreda Road near Belfast. Rather than taking the car, farmers could have gone to their nearest railway station and taken the train for central Belfast. From here, they could have taken a short walk to the City Hall where the No 8 trolley bus would take them to the Cregagh terminus in East Belfast. Mr McCormick's farm was now, just a three minute walk away.

Before reaching the farmyard at Willow Bank, the visitor may have seen some fine examples of the Ayrshire breed grazing the surrounding pastures. This herd, founded with some of the best strains of both Irish and Scottish breeding, featured several prize-winners. Although most cattle would have been pedigree Ayrshires at Willow Bank during the mid 1940s, there were also, according to old farm records, one or two pedigree Jerseys and a few crosses between the Jersey and Ayrshire breeds.

Supposing a prospective heifer-buyer had visited the farm in September, 1946, they would have seen some fine young red and whites in the form of 'Willow Bank Fiona' (born 20 September 1944), 'Willow Bank Daphne' (born 6 September 1944) and 'Willow Bank Iris' (born 29 September 1944). The heifers were all out of bought-in cows bred by Mr David Wright of the Moat, Dundonald (UFU secretary), John Andrews and Co Ltd, Maxwell Court,

Comber, and George Morton, Newton-Stewart, Wigtownshire, respectively.

Having had the opportunity to run their eye over each of the three heifers, the visitor addressing Mr McCormick may have said, "I'll give you ninety pounds apiece". The response to this offer may have been, "Those ones are not for private sale, if you want them go to Ayrshire breed sale at Robson's mart next week". Of course, this is all supposition. It is a fact, however, that 'Willow Bank Fiona', 'Willow Bank Daphne' and 'Willow Bank Iris' were all paraded at the Irish Ayrshire Cattle Society's third annual sale, which took place on Friday 25 October 1946.

This was not the only opportunity made available for those farmers wanting to buy pedigree Ayrshire cattle during that year. Several on-farm dispersal and reduction sales took place in 1946.

Having sold his farm, Mr Thomas McCann of Droagh put his 'Parkmount' herd up for public competition at an auction conducted by JA McClelland & Sons, Ballyclare. It comprised forty good commercial cows and sixteen pedigree Ayrshires, including the three-year-old stock bull 'Tildarg Bruce'.

Several weeks later, in March 1946, it was the turn of Mr David Dodds of Main Street, Saintfield, to reduce the numbers of his Gold Cup winning 'Glebeland' herd that had been established around 1921. This sale comprised 11 cows, 6 newly calved heifers, 4 yearling heifers, 3 heifer calves, a young bull and the herd's Stock bull.

In September 1946, the unreserved clearance of the 'Barn Mills' herd took place following the sale of the flax spinning mill and a major proportion of land, owned by Messrs. James Taylor & Sons Ltd. Bottles of Ayrshire milk had been sold from their farm which was situated

off the Belfast to Whitehead Road at the eastern end of Carrickfergus (Taylor's Avenue).

This dispersal, conducted by JA McClelland on Thursday 26 September 1946, featured 19 cows in-calf to Barn Mills Blakeney, 9 heifers (between sixteen and twenty month of age), 10 heifer calves, 6 bull calves and the stock bull. One month after the Barn Mills dispersal, it was time for the Irish Ayrshire Cattle Society's aforementioned third annual show and sale, held in Robson's yard, Belfast.

The first Irish Ayrshire Cattle Society Show and Sale took place during the Second World War, the second one almost six months after the end of hostilities and the third such event in the same month that the Nuremberg trials in Germany came to an end. Indeed, just nine days before the show and sale in Belfast, the execution of Goering, Ribbentrop, Keitel and eight other war criminals had been reported.

On a lighter note, that was also the same month the Duchess of Windsor made her first return visit to England following her marriage to the one-time King Edward VIII. Perhaps of greater interest to *Farm Week* readers is that during the same month that the sale took place, Mr Cole, Unionist MP for East Belfast, raised the matter of a lack of tractor spares in the House of Commons and Mr Harry Ferguson announced the "latest Ferguson accessories would be made in Belfast" and on a wider scale "forecast a brighter future for the farmers of the world".

Perhaps, this healthy attitude of optimism was evident ringside when the bidding got underway at the Ayrshire show and sale, which featured the cream in terms of Ayrshire genetics, not just from Ireland but from the breed's homeland as well. Indeed, in an attempt to encourage two hundred plus Ayrshire Society members from Northern Ireland, one of the

Scottish breeders put up a calf, the proceeds from the sale of which were to be given to the revived Irish Club. The calf fetched 80 guineas, paid by Mr Charles J Webb of Randalstown. *Farm Week* readers may be wondering how Mr William McCormick's three heifers faired when brought into the sale-ring. Details on the prices they realised, and many of the other lots forward at this 1946 Ayrshire show and sale, are published below.

Prices realised at the Irish Ayrshire Cattle Society's third annual show and sale held in Robson's mart, Stewart Street, Belfast on Friday 25th October 1946.

Cows: Hillend Beauty (bred by David Howie, Crossford, Lancashire) to J Dunlop, Larne for 170 gns, Smithston Idris (James H McClymont, Patna, Ayrshire) to J Dunlop, Larne for 150 gns, Willow Bank Fiona (Mr McCormick, Willow Bank Upper Knockbreda Road, Belfast) to AH Millar, Straidarran, Co Londonderry for 125 gns, Crankill Violet (James Russell, Crankill, Glarryford) to T Craig for 115 gns, Willow Bank Dahne (Mr McCormick, Willow Bank, Upper Knockbreda Road, Belfast) to L Mackie, Lisanoure Farms Ltd, Loughguile, County Antrim for 105 gns, Lismanery Delight (Samuel McCrone, Ballynure, Ballyclare) to J Dunlop, Larne for 100 gns. Hillend Chocolate Cream (David Howie, Hillend, Lanarkshire) to J Livingstone for 100 gns. Millview Maise (William Davidson, Spa Poultry Farm, Ballynahinch) to M Kyle for 95 gns. Willow Bank Iris (Mr McCormick, Willow Bank, Upper Knockbreda Road, Belfast) to RO Hermon JP, Necarne Castle, Irvinestown for 94 gns. Gobbin Hill Patsy (Nelson W Kane, Granshaw, Islandmagee) to TJ McMillan for 92 gns. Cauldcoats Moira (Alexander White, Linlithgow, West Lothian) to AH Millar for 91 gns. Lismanery White Rose (Samuel McCrone, Ballynure, Ballyclare) for 80 gns. Loaninghead Mirander (James Turner, Balfron Station, Glasgow) for 75 gns. Chapelhill Bessie (Walter Biggar 7 Son, Grange Farm, Castle Douglas) to Major TJ Graham for 67 gns. Clandeboye Violet (William E Ardill, Ballysallagh House, Clandeboye) to HA Porter for 67 gns. Killagan Belle (John H Kissack, Killagen, County Antrim) to J Connor for 66 gns. Gilhar Snow White (AF Gilmour, Faulds, Renfrewshire) for 65 gns. Cairnbrook Iney (David Rennie, Little Cairnbrook, Stranraer) to H J O'Reilly for 58 gns. Ballygelly Madge (Robert McMaster, Roughan, Broughshane) to AG Bailie for 55 gns. Willow Bank Eva (Mr McCormick, Willow Bank, Upper Knockbreda Road, Belfast) to W Gordon for 54 gns. Thorn Kate (William Lorimer, Thorn Cottage, Crumlin) to W Gordon for 51 guineas.

Bulls: Carnell Kenneth (George Templeton, Hurlford, Kilmarnock) to RD Turkington, Lurgan for 75 gns. Rathfern Merry Boy (L Mackie, Rathfern, Belfast) to Mr Dunlop for 74 gns. Ballynadrenta Super Grade (Thomas J Suffern, Crumlin) to Dick Bros. Ballynadrenta Radio (Thomas J Suffern, Crumlin) to Antrim Mental Hospital for 66 gns. Willow Bank Eclipse (Mr McCormick, Willow Bank, Upper Knockbreda Road, Belfast) to E Connor for 62 gns. Ballyclose Brigadier (WA Harbison, Ballyclose, Cullybackey) to J Black for 50 gns. Hillsborough Steadfast (Hillsborough Agricultural Research Institute, Hillsborough) to J Gibson for 50 gns. Glenview Edmund (George Crawford, Glenview, Castlereagh, Belfast) to Jas Smyth for 50 gns. Fenaghy Young Showman (William L Young, Fenaghy, Cullybackey) to H Creaney for 46 gns. Crankhill Supreme (James Russell, Crankhill, Glarryford for 45 gns. Movilla Dandy (George Wallace, Movilla House, Newtownards) to J Johnston for 42 gns.

Farming exploits of Robinson and Cleaver director

> *"Different people would have had different reasons for remembering the 1934 Balmoral Show".*

At the Balmoral Spring show of 1934 the King's Hall was officially opened by HRH The Duke of Gloucester and Edward A Robinson of Terrace Hill exhibited some of his foundation Jersey cattle.

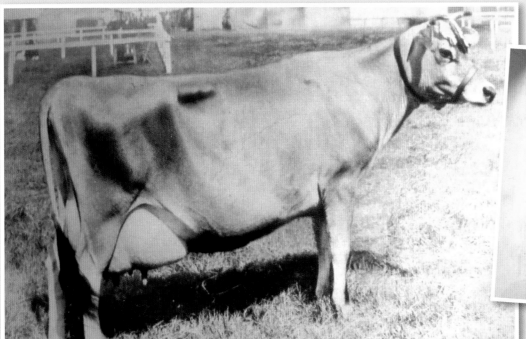

Over 60 years ago this fine Jersey cow would have been an excellent example of the breed.

This bull box and pens for young stock were occupied by Mr EA Robinson's pedigree Jersey cattle at Terrace Hill.

Different people would have had different reasons for remembering the 1934 Balmoral Show. For example, for the Royal Ulster Agricultural Society's Secretary/manager, Mr Sam Clark, that was the year the newly built 'King's Hall' was opened. For His Royal Highness the Duke of Gloucester, that was the year he officially opened the 'King's Hall' and was given an overview of Northern Ireland's most important industry ... Agriculture. At this time there were around 70,000 farm holdings across the Province.

For the well-known Belfast businessman Mr Edward A Robinson of Terrace Hill, the 1934 Balmoral would also have been memorable, not so much in connection with the opening of the King's Hall but rather because it was at that show his pedigree Jersey cattle made their debut on the cattle lawns.

Perhaps, it would have been the Royal Ulster Agricultural Society's Mr Sam Clark who would have been under the greatest pressure during the days running up to the show. As he came into work each morning and was met with the sight of many workman and a partially finished building, his over-riding thought may well have been, 'Will the King's Hall be ready in time for this year's show?' Ever since local demolishers R & W Lusby had removed the old temporary building that fronted the showground, the 'clock had been ticking'. To erect a building which would be Ulster's equivalent to London's 'Olympia Hall' was an immense project.

After a final push by the contractors and with the paintwork barely dry, the new building was ready and as Mr Clark watched the Duke of Gloucester KG, GCVO, make his way through the mahogany panelled entrance hall and across

the marble terrazzo floor, he may have breathed a huge sigh of relief. Although the show was not over, one of the main hurdles for that 1934 event had been overcome.

For Mr Edward A Robinson this 1934 show would give him his first 'Balmoral experience' as a cattle exhibitor. Although Mr Robinson owned a farm beside his large house at Terrace Hill near Shaw's Bridge on the outskirts of Belfast, he was better known as a director of 'Robinson and Cleaver'. Indeed, some of *Farm Week*'s more senior readers may recall being taken to this fine department store during infrequent visits to Belfast City.

Mr Robinson, it would seem, was a man of diverse hobbies, including dogs, racing pigeons and thoroughbred horses. Alsatians and Irish terriers were his breeds of choice within the canine species, while birds from his large loft

The back door of 'Wotton Said's bull box opened up onto these pastures at Terracehill.

won, on at least two occasions, the 'King's Cup. Regarding horses, perhaps Mr Robinson's most famous one was a filly called 'Gainsworth', which won the Irish 1,000 guinea flat race at the Curragh in County Kildare. Readers of *Farm Week* will, perhaps, want less information on shopping, dogs, pigeons or racehorses and more about Mr Robinson's pedigree Jersey cattle that were registered in the English Jersey Cattle Society (EJCS) under the prefix 'Terrace Hill'.

Following the old adage 'If you are going to keep cattle you might as well keep good ones', Mr Robinson selected his foundation animals from one of England's top Jersey herds owned by Mrs Evelyn of Wotton House, Dorking. According the EJCS herd book for 1932, Mrs Evelyn, a past President of the Society, was milking around 45 Jerseys, many of which had qualified for a register of merit. At this time the herd included a 14-year-old wartime-born matron called 'Wotton Maid of the Mist' and several top show cows. That year Wotton cows had won prizes at the English Royal, Sussex, Tunbridge Wells, Oxfordshire, Bath and West, Royal Counties, Tring and London Dairy Shows and several animals were exported to Portugal and South Africa.

Of the seventeen calves born at Wotton

BREAKDOWN OF MR EDWARD A ROBINSON'S TERRACE HILL HERD OF PEDIGREE JERSEY CATTLE DURING THE YEAR OF 1938

BULLS AND BULL CALVES

Wotton Said	Born 8.7.1932
Terrace Hill Best Boy	Born 24.4.1936
Terrace Hill Baron	Born 12.6.1937
Terrace Hill Punch	Born 24.5.1938 (calf)
Terrace Hill Rest	Born 3.9.1938 (calf)

COWS AND HEIFERS IN MILK

Bashful Ruby	Born 2.5.1933
Marston Retry	Born 23.1.1934
Terrace Hill Betsy	Born 8.5.1934
Marston Polly	Born 20.1.1936
Terrace Hill Judy	Born 20.3.1936
Terrace Hill Ruby	Born 25.7.1936

HEIFERS AND YOUNG FEMALE STOCK

Terrace Hill Oxygen	Born 13.7.1936
Terrace Hill Pink	Born 1.5.1938 (calf)
Terrace Hill Polly	Born 31.5.1938 (calf)
Terrace Hill Ruby 2nd	Born 8.8.1938 (calf)

House during that year of 1932 three (two heifers and a bull) were destined to come across the Irish Sea to lay a foundation to Mr Edward Robinson's herd at Terrace Hill. This took place during February 1934. Their names were 'Wotton Air Link' (heifer born 10 June 1932), 'Wotton Oxygen 2nd' (heifer, born 30 December 1932) and 'Wotton Said' (bull, born 8 July 1932). These animals were yearlings at the time of importation, and they were accompanied by a young cow called 'Wotton Bersinda'.

Three months after these animals had been transferred, Mr Robinson made his debut at the aforementioned 1934 show. Also exhibiting Jersey cattle at this historic show were The Right Hon J Milne Barbour, DL, MP, Conway, Dunmurry; the executors of WH Odlum, Ardmore, Co Wicklow, Mr James C McGifford, Cromlyn Lodge, Hillsborough; Mr WH Richardson, High Street, Portadown; Mr John McCaldin, Knockroe, Monaghan; Mrs Isabel McClenehen, Rathfriland; and Mr James E Anderson, Dunmurry.

Just as each one of these Jersey breeders won prizes with their cattle in the Jersey classes at the 1934 show, so too did Mr Edward Robinson with three of his foundation beasts. 'Wotton Betsinda RM' took a first and EJCS Special in the cow class, while the heifers 'Wotton Oxygen 2nd' and 'Wotton Air Link' took a first and third in their respective classes. Thirteen Jersey bulls were turned out before the judge L Gordon Tubbs and Mr Robinson's rising two-year-old, 'Wotton Said', took a third ticket in his class. As a three-year-old and four-year-old 'Wotton Said' was back at Balmoral when he lifted the top honours on both occasions.

Back on his home farm, 'Wotton Said' was kept in a bull-box, the back door of which opened up to a picturesque paddock. Today, this building and the main byre that once housed Mr Robinson's Jerseys (locally known as Ed's cows) largely remain the way they were seventy-five years ago.

Between 1934 and 1947 Mr Edward Robinson continued to register his Jersey cattle in the Society's herd book. During 1947 two great-grandsons of 'Wotton Said' were born in the herd and duly registered under the names 'Terrace Hill Visto' and 'Terrace Hill Dandy'. On Sunday 7 December Mr Edward Arthur Robinson JP passed away at his Terrace Hill residence.

"animals were not selected on fancy show points or for colour, shape or size but rather on their potential to produce butter."

Breeding a Jersey herd for its butter qualities

Although generally not renowned for their mild temperament, Jersey bulls have been kept on Ulster farms over the decades to sire daughters with greater milking potential.

Those consumers wanting to put something that little bit special over their breakfast cereal or in their morning coffee can do so by adding a splash of milk from Jersey cows. Today, this special product has filled a niche in the marketplace and as such can be found on the shelves of select supermarkets. Delicious ice-cream made from the Jersey milk is also available, but what about butter? At present, those people wanting to spread real 'Jersey' butter over hot toast or use real 'Jersey' butter as an ingredient for cream cakes may find it a little more difficult to source.

A century ago, one of England's top 'Butter' herds of Jersey cattle was owned by Dr Herbert Watney of Buckhold, Pangbourne, and Reading. Without having any great knowledge about breeding cattle, Dr Watney founded his herd around 1885 with that year's English Jersey Herd Book giving details on some twenty-three bought-in beasts. 'Jealous Maid', 'Pretty Lass', 'Snowdrop 2nd', 'Silence' and 'Japonica 3rd', were included in the list, some of the cattle having been bred on the island of Jersey and others on English farms.

All these foundation animals were not selected on fancy show points or for colour, shape or size but rather on their potential to produce butter. From the outset Dr Watney was singularly focussed and carefully

OFFICIAL MILK RECORDS IN 1927

Name of cow	Age in years	Days in milk	Yield in lbs	% butterfat	Yield in butterfat	Class of register merit
Sylvia's Pet	6	349	10,962	5.2	569.86	A
Marcella's Gem	6	343	8,540	4.9	418.6	C
Lady Erin	4	332	9,464	5.07	479.81	A
Irish Lassie	3	324	7,571	5.09	385.23	B
Lady Times	6	349	11,213	4.42	495.87	B
Lady Gem	4	361	10,908	5.55	605.51	-
Irish Times	6	336	11,174	5.37	600.30	A
Spangolite	4	321	9,934	5.24	520.31	A
		Average	9,971	5.11	509.45	

monitored the butter-producing capability of each cow to identify those extra-ordinary animals which could be retained for breeding. The results were phenomenal. According to Professor Robert Wallace (Farm Livestock of Great Britain) the homebred Watney animals had developed powers of butter production beyond anything that money could buy.

By the turn of the 20th century his herd was winning more gold medals in the butter classes than any other registered in the English Jersey Cattle Society herd book. Around this time a group of researchers who were compiling an Encyclopaedia of Agriculture visited Dr Watney's Buckhold herd to seek his opinions on Jersey cattle breeding. After a series of conversations and a tour around the farm, they recorded that Dr Watney was a "Master of the subject" and went on to extol his methods.

By keeping his cows healthy, warm and well-fed, Dr Watney maximised their butter production. In summer, these animals were not only grazing knee-deep on lush pasture but were given a daily allowance of corn. When inside during the colder months Buckhold cows were given as much good hay as they wanted, served alongside cattle-cake, parsnips and cabbage.

It is not surprising that with such 'five-star treatment' Dr Watney's cows delivered. Information obtained from his farm sales book shows that between 1 October 1916 and 30 September 1917 over 13,000 lbs (5897 kgs) of butter had been produced. One of the cows that contributed to this impressive figure was called 'Aurelius's Time' and she, *Farm Week* readers may be interested to know, was the mother of a five month old bull which crossed the Irish Sea to join a Northern Ireland herd. This young animal had been purchased at one of Dr Watney's draft sales, held during the First World War.

This event, which took place on the home farm on Wednesday 18 July 1917, featured forty-five cattle made up of twenty-six females and nineteen males. The cows and heifers averaged £46 17s 8d and the bulls £16 5s 0d. The aforementioned bull that came to a County Tyrone herd was called 'Times Gem' and he was purchased by Dr James Scott Gordon DSc, CBE of Stragollen, Strabane, for ten guineas.

Although the herd was based in County Tyrone, during the 1920s all correspondence was sent to Beaconsfield on the Knock Road, Belfast. This was because in addition to being a farmer, Dr Gordon was the first Permanent Secretary to the Northern Ireland Minister of Agriculture. When the local government was set up at Stormont, Minister Sir Edward Archdale chose Mr Gordon for the post because of his deep knowledge of Agriculture. During his service as Secretary, which exceeded ten years, Mr Gordon brought great enthusiasm and a pioneering dynamism to the job. He played a major role in the establishment and implementation of the Northern Ireland Marketing of Eggs Act and, in latter years, the Pigs Marketing Board.

While Dr Gordon would have been able to speak with authority on a wide range of agricultural matters perhaps, when it came to the subject of Jersey cattle he would have been in his element. His select herd was milk recorded and this at a time when many Ulster farmers were working through their local milk recording associations, to weed out those '2 % boarders', so unworthy of their keep. While it was said one could have seen an old penny lying on the bottom of a pail full of milk from such cows, this would not have been the case with those of the Jersey breed.

When 'Times Gem' was reared up and put to work in Dr Gordon's herd he did what was expected and up to the time of being sent to the butcher at six years-of-age in June 1923, he sired some good daughters, one of which was entered in the milking trials at the Royal Ulster Agricultural Society's Balmoral show in 1927. Her name was 'Sylvia's Pet', calved 11 May 1920. Dr Gordon had bred her dam 'Sylvia' (calved 17 March 1917) and grand dam 'Spangle' (calved 26 March 1912).

Four hundred and forty-five cattle were entered at this 1927 Balmoral show, which just under thirty-five thousand people attended over the first three days. Not only did the show provide the farming community with an opportunity to socialise, it also provided the Northern Ireland Ministry with a platform on which to bring in fresh thinking. Demonstrations at the event covered such topics as 'Clean Milk Production', 'Chicken Trussing', 'Packing and Grading eggs' and the 'Dishorning of Calves'.

In addition to the inspection classes for the various species and breeds of farm livestock, milking trials were laid on for both goats and cattle with Mr George S Robertson of Queen's University in attendance as official tester. For the aforementioned Jersey breeder and first Permanent Secretary to the Ministry of Agriculture, these milking trials brought success when 'Sylvia's Pet' took a first prize, a certificate of merit and a gold medal, having given in one day 40.46 lbs of milk and 2.38 lbs of butterfat.

The cow was awarded the 1st prize in the milking trials, a certificate of Merit and a Gold Medal for highest yield butterfat open to all breeds. Official milk records for 'Sylvia's Pet' and other pedigree Jersey cows are printed in the accompanying table.

FOR SALE

PEDIGREE JERSEY BULL

Nine-months-old; excellent milking strain; dam gave 828 gallons as a two-year-old; sire's grand-dam gave 1,261 gallons.

For particulars apply to:

JS Gordon, Beaconsfield, Knock Road, Belfast.

Reproduction of 1925 sales notice.

Over the years many Jersey cattle from JS Gordon's pedigree herd were purchased by farmers all over Ireland. A few went across the water to work on English and Scottish farms. After a distinguished career as an agriculturalist, James Scott Gordon, BSc, CBE passed away at his Stragollen Residence during September 1946. He was in his 79th year and was buried in Strabane Cemetery.

"The whole is greater than the sum of all its parts and so it must have been with a great sense of optimism that the members of the newly formed 'Northern Ireland Galloway Cattle Association' took their leave"

Galloways strong in the early part of last century

The year is 1905 and passing through the pleasant County Antrim village of Ballynure and on to the higher lands of Ballyboley, the traveller will soon arrive at a particular farm which is home to one of the best pedigree Galloway herds in the north of Ireland. Fourteen society registered calves have been born on that holding during this year, some of which had parents or grandparents from the breed's top prefixes including 'Castlemilk' (owned by Sir RW Buchanan-Jardine, Bart of Castlemilk, Locherbie), 'Stepford' (owned by David Brown, Stepford, Auldgirth, Dumfries) and 'Craighouse' (owned by W&D Wilson, Craighouse, Lockerbie).

Passing alongside the pastures at Ballynure during the month of July 1905, the sturdy young bull calf seen grazing alongside his dam may have been 'Pat of Ballyboley' (9615). He had been dropped the previous February. Other bull calves to have been seen on this holding at the time would have been 'Antrim' (c. 2 January 1905) and 'Ulster' (c. 7 January 1905).

This prizewinning Galloway bull was photographed around 100 years ago.

Eleven Galloway heifer calves were also born on this farm during that year and their pedigrees, along with the three bulls, were published in the Galloway Cattle Society's twenty-sixth herd book. In these herd books their linage could be traced back to great foundation bulls including 'Sir Walter' (536), 'Wellington' (22) and 'The Squire' (18). Back in 1877 these animals had been entered in the first herd book that was edited by the Very Rev John Gillespie, LLD, Mouswald Manse, Ruthwell, Dumfriesshire. This gentleman, a Parish Minister for some forty-seven years, had acquired his love of Galloway cattle from his father.

In addition to editing the Galloway breed Society's herd book year on year, Rev Gillespie also played an influential role in agriculture, being largely responsible for founding the West of Scotland Agricultural College. He had, according to the authors of the book *Great Farmers*, a natural dignity and a great gift for making friends. Such was Rev Gillespie's influence that he was unofficially known as the 'Minister of Agriculture for Scotland'!

Rev Gillespie was an accomplished writer on a wide range of agricultural matters and his work was published in various journals and books including the *The Breeds of Livestock* (1916) and *North British Agriculturist*. Although Rev Gillespie wrote articles on such topics as 'Farmers and Income Tax', it can be safely assumed his favourite subject, with regard to agriculture, was Galloway cattle. Writing in the latter journal, dated 25 November 1891, he made the following comments on the Galloway beast's coat and skin:

"The skin and hair were outstanding qualities of the Galloway breed. The skin should be moderately thick, but always mellow and soft. The hair should of course be black, but not of a jet or inky black, the breeders liked to see a nice brown tinge, and all the best bred specimens would be found to have that characteristic". "Galloways", he continued, "had always two coats of hair fully developed. The upper coat should be long and soft, but not too curly. Animals with a very curly coat were generally indifferent thrivers. The under coat should always be soft, close and silky. This was a very important point in the breed, because

Sapere Aude – Dare to Know. This crest, signed by Robert I Calwell (1854–1927) was found in his own Galloway Cattle Herd Book for 1906.

the animal's ability to resist cold and stand exposure depended on it."

Given the breed's inherent hardiness it's highly unlikely that during a fall of snow, those aforementioned Galloway cattle being out-wintered at Ballyboley would have been stampeding for the hedgerows. At this point, some *Farm Week* readers may be wondering who owned this particular pedigree herd, which was, at the time, one of the largest in the north of Ireland. It was owned by a gentleman called Mr Robert I Calwell CE.

Born in 1854, Robert Calwell was the second son of Mr Robert Calwell of Annadale, County Down. Following his education at the Royal Academical Institution and in Paris, Robert attended the old Queen's University from where he obtained the degree of 'Bachelor of Civil Engineering' in 1874. Having set up an architectural practice in Belfast, Robert Calwell started to work closely with the city's Corporation, serving for a time as City Surveyor and Chairman of the panel of architects working on the Corporation's Housing Scheme. Mr Calwell also served as Civil Engineer to the Belfast Central Railway prior to its purchase by the Great Northern Railway. In recognition of his services, Mr Calwell was made a Commander of the Order of the British Empire in 1921.

Away from the bustle of Belfast City and his offices at 16 Donegal Square, South, Mr Calwell would, at times, have been seen with his pedigree Galloways at Ballyboley. Although he enjoyed fishing, shooting and owned several racehorses, the breeding of pedigree cattle was Mr Calwell's chief hobby. His large herd was 'professionally' managed and, over two decades, was frequently represented at the Royal Ulster Agricultural Society's Balmoral Show. Indeed, one hundred years ago at the 1908 event, 'Minx 5th of Ballyboley' took the red badge in the class for Galloway Heifer calved on or after 1 December 1905.

Mr Calwell's name last appeared in the Balmoral prize-winner's list twenty years later, at the 1925 show, when four of his cattle won prizes. 'Rabbi of Ballyboley' won a first in the 'Galloway bull, any age' class, while in the class for 'Galloway cow or heifer, any age' the first, second, and third tickets went to 'Laura of

Ballyboley', 'Ruth of Ballyboley' and 'Outshine of Ballyboley', respectively.

On 26 January 1927, Mr Robert I Calwell passed away at his residence, Carnivard, Annadale, County Down, following an attack of influenza. Less than three weeks later a significant event in terms of the Province's Galloway history took place, when a dedicated breed sale took place at a venue chosen by the Ministry of Agriculture in that delightful County Antrim village of Cushendall.

This was the first such sale of its kind and attracted widespread interest as farmers flocked keen to secure the type of hardy beast that would thrive in all weather conditions on the surrounding mountains and glens. Shifting the various lots would not have presented the auctioneers, RJ Allam Ltd, with an 'uphill struggle', especially when it came to the sale of two particular bulls that were sons of the Highland Show Champion, 'Barmark Eclipse'.

These animals, having taken the first and second places in the show, fetched 51 guineas and 33 gns in the sale, having been purchased by Captain Casement DSO and Mr David Cupples of Clough, County Antrim. A show and sale report stated that these fine bulls had been exhibited by a "new exponent to the Galloway breed", Mr Isaac C Hay of Beechview, Bellahill, having been "bred in the herd of the late Mr Robert I Calwell".

After the sale, a group of farmers gathered with a mind to promoting the Galloway breed, with Mr RJ Allam presiding. After his opening remarks, steps

This prizewinning Galloway cow was pictured around 100 years ago.

were taken to set up a new organisation with Captain Casement DSO being elected to the Chair and Mr James C McElheron, Altacoan, Cushendall to the important role of honorary Secretary. A committee was made up from remaining breeders. The whole is greater than the sum of all its parts and so it must have been with a great sense of optimism that the members of the newly formed 'Northern Ireland Galloway Cattle Association' took their leave of McAlister's mart back on that winter's day of 14 February 1927.

"Inscribed were the words, 'Presented to Sir George Macpherson-Grant of Ballindalloch from breeders and others interested in Aberdeen Angus Cattle, in recognition of his eminent services in improving the breed.'"

'The Pearl of the North' a real jewel in the crown for visitors

The Ballindalloch Castle Estate, Bannffshire, Scotland, is home to the world's oldest herd of Aberdeen Angus Cattle.

One doesn't need to have a particular interest in Aberdeen Angus cattle to enjoy a visit to the stunning 16th century Ballindalloch Castle, set in the magnificent surroundings of the Spey Valley in the North of Scotland. Absolutely not, for those people with an appreciation of beautiful architecture, history, culture and artistry will benefit so much from a visit to this wonderful Highland home to the Lairds of Ballindalloch. It is not surprising that this Castle is known as the 'The Pearl of the North'.

The grounds in which Ballindalloch Castle is set are maintained in pristine condition. There is a beautiful walled garden where visitors, at this time of year, can appreciate the lovely 'Ballindalloch Rose' along with other blooms to the relaxing sounds and sights of several impressive water fountains. A stroll along the tree-lined paths brings one to the Grass

Labyrinth, expansive rockery or golf course. In Scotland pigeons are sometimes called 'doos' and there is a large Doocot (dated 1696 with 844 stone nesting boxes) to be seen on the Ballindalloch Estate.

Having undertaken a walk round the estate's grounds, visitors can 'refresh the tissues' in the Castle's tearoom. Once here, a certain hesitation can creep in, especially regarding the homemade cakes. After the fresh scones, should the jam and cream sponges be sampled or is the chocolate cake more tempting, bearing in mind that also on display is a selection of mouth-watering tray bakes. The Ballindalloch tea house can indeed prove a testing place to the indecisive.

Having drained the coffee cup and finished the last crumbs of cake, a tour inside Ballindalloch Castle cannot fail to impress. There is the Hall with its grand staircase, the Drawing Room laid out in 18th century furniture, the 19th and 20th century toys inside the Nursery and the library holding around 2,500 volumes. The large Dining Room leaves it mark as well, being panelled in American pine and adorned with exquisite tapestries and paintings.

But it is the Silver Cups on display in this room that may be of special interest to visiting farmers. These had been won by Sir George Macpherson-Grant (1839–1907), 3rd Baronet for cattle from his world-famous herd of Aberdeen Angus cattle. The 'Ballindalloch' herd dates back to 1860 and today it continues to flourish under the stewardship of the founder's descendant, Mrs Macpherson-Grant Russell. During the grazing season, some of the present Ballindalloch herd members can be seen on an ancient pasture referred to as the 'Cow Haugh' situated beside the Castle.

When writing about the founding of the Aberdeen Angus breed in their book *Great Farmers*, authors James A Scott Watson and May Elliot Hobbs make special mention of three men. The first two were Hugh Watson of Keillor in Forfarshire (born 1789, died 1865) and William McCombie of Tillyfour (born 1805, died 1880) with the third 'triumvirate' being the aforementioned Ballindalloch herd founder Sir George Macpherson.

It was in 1860 that Sir George purchased a cow named 'Erica' for fifty guineas from James Carnegie, 9th Earl of Southesk (1827–1905), owner of Kinnaird Castle that nestled between the Angus coastline and glens of Northern Scotland. 'Erica' (843) was by 'Cupbearer' (59), a grandson of 'Old Jock' (1), and herd dam 'Emily' (332) was also by 'Old Jock' (1). She was described as being "not a very large cow, but standing on very short legs and having a lovely feminine head and splendid quality".

Erica was the dam of four daughters 'Erica 2nd' (1284), 'Eisa' (997), 'Enchantress' (981) and 'Elba' (1205), and it was from these four cows that the most notable line was developed at Ballindalloch. Writing in 1910 the authors Macdonald and Sinclair recorded that "No family of polled cattle had in recent years taken a more distinguished position in the show yard than the Ballindalloch Ericas". Regarding the show bulls raised at Ballindalloch, one called 'Judge' stands out, as in 1878 he won a first prize at a Paris Exhibition.

A group of fellow Aberdeen Angus breeders in Scotland decided to acknowledge the work of Sir George in 1904. One day, they gathered at Ballindalloch and presented him with a large silver tankard. Inscribed were the words, 'Presented to Sir George Macpherson-Grant of Ballindalloch from breeders and others interested in Aberdeen Angus Cattle, in recognition of his eminent services in improving the breed'.

One of the several pedigree Aberdeen Angus herds in the North of Ireland that benefitted from an infusion of 'Ballindalloch blood' was that of the 'Lisnabreeny' prefix, established by Frederick Joseph Robb MA LLB near Belfast in 1898. Mr Robb (born 22 September 1869) was a barrister in the firm of John Robb & Co Ltd, Belfast. For many years there was rarely a Balmoral or Dublin show that did not feature a top prizewinning 'Lisnabreeny' beast.

Following the death of Mr FJ Robb on 22 April 1922, his nephew, Dr Alfred Arthur Robb DSc, PHD, FRS, took over the Lisnabreeny herd. Although widely known for this scientific research concerning relativity, Dr Robb was, perhaps, best known in Ulster's farming community as an Aberdeen Angus breeder. During his years at the helm of the Lisnabreeny prefix the 'Ballindalloch' influence continued through a bought-in bull called 'George R of Ballindalloch' 30611.

In addition to the pedigree Aberdeen Angus cattle, old farm sales notices reveal that there were also four milch cows and twenty-five head of laying hens on Dr Robb's farm at Lisnabreeney House, Belfast, during the mid to late 1930s. Horse-drawn implements and miscellaneous items on the holding at this time included a 'McCormick' Reaper, 'Albion' Self-binder, three horse cultivator, Double Drill Turnip sower, Horse Hay Rake, Two Rick-Shifters, Two two-horse Grubbers, Rubber-tyred Governess Cart, Two Cattle Floats, Two farm Carts, and a Turnip pulper.

Dr Robb's last distinction with an Aberdeen Angus beast took place at the Royal Ulster Agricultural Show of 1936 when his heifer, 'Euthalla 3rd' of Ballintomb, won the female championship. On 14 December that same year Dr Robb passed away. Subsequently, the famous Lisnabreeny herd of pedigree Aberdeen Angus cattle was dispersed at the Balmoral Cattle Sales on Wednesday 24 February 1937.

Warning sign on a side gate to the 'Cow Haugh' at Ballindalloch Castle.

ADMISSION FREE.
THE BULL
WILL CHARGE
YOU LATER!

A group of Ballindalloch Aberdeen Angus cattle grazing in the 'Cow Haugh' situated beside the castle.

The herds of Cambridge Grubb are put under the spotlight

The sale notice for Mrs Cambridge Grubb's herds of Aberdeen Angus cattle and Large White pigs which were dispersed in Belfast on Friday 22 November 1901.

Some readers of *Farm Week* will remember the television programme that brought viewers into the home of well-known personalities. Their identity was to be deduced as the camera moved from room to room and many clues could be found by studying the personal possessions of the mystery householder.

For instance, if the camera revealed a pair of plimsolls and a tennis racket, the conclusion that the celebrity had a sporting nature could be made, whereas a double-barrelled gun and a glass case containing a couple of stuffed partridges, would indicate an interest in countryside pursuits.

If, around seventy years ago, a camera had been brought into a particular house near the main road between Lisburn and Dunmurry, viewers could have attempted to guess the name of a well-known lady who was a successful breeder of pedigree livestock. As the camera rolled, the following scene may have come into focus.

At the turn of the 19th century, some fine pedigree Aberdeen Angus cattle could have been seen in Mrs Cambridge Grubb's County Antrim herd at Killeaton House, which was situated near the main road between Lisburn and Dunmurry.

SALE TO-MORROW.
DISPERSAL SALE.
HIGHLY-IMPORTANT SALE BY AUCTION
OF
PURE-BRED
ABERDEEN-ANGUS CATTLE.
MESSRS. ROBSON
Have received instructions from Mrs. Cambridge Grubb, Killeaton House, Dunmurry, County Antrim,
TO SELL BY AUCTION, IN THE
ROYAL VICTORIA HORSE BAZAAR,
31 and 33, CHICHESTER STREET, Belfast,
On FRIDAY, 22nd November, 1901,
At One o'clock,
HER WELL-KNOWN PRIZE-WINNING HERD of ABERDEEN-ANGUS CATTLE, consisting of Cows, Heifers, Calves, and Bulls.
The above Herd has been most successful in prize-winning in the principal Shows, and the Sale presents a rare opportunity of securing animals of the purest blood in the kingdom.
Also, TWO PURE-BRED LARGE YORKSHIRE SOWS AND BOAR.
Full particulars and Pedigrees in Catalogues, which can be had from Messrs. ROBSON.
Further Entries of Pedigree Stock solicited for this Sale.
27742

On entering the fine three bay, two-storey mansion (built around 1870), one would have been immediately impressed with the high quality of the furniture. No expense had been spared in the acquisition of many fine pieces, a Chippendale Mahogany table with six matching Fish Tail Chairs, the Georgian Mahogany sideboard and oak court cupboard carved with the inscription 'THM1670'. There were many books in the room and adorning one of the main walls, a large oil painting of King William III alongside a set of Ivory Tusks!

Because all these items were mentioned in a house clearance sale we know that they could have been seen, around 1940, in this particular house. By looking at old records in connection with the property, it would not have been surprising if the camera had picked out some fading photographs of prize cattle. Having surveyed all this information, local people would have little difficulty in identifying the 'mystery lady' as Mrs Cambridge Grubb of Killeaton House, Dunmurry.

Mrs Grubb, who married Mr Richard Cambridge Grubb in 1870, was a daughter of Mr Jonathan Richardson JP of Glenmore House, Lambeg, County Antrim. Although her father was a breeder of pedigree Shorthorn cattle (he purchased a bull called 'Baron Warlaby' from the famous Richard Booth in the 1830) it was to the Aberdeen Angus breed that Mrs Grubb turned when establishing her own pedigree herd in 1894.

Travellers on the main Lisburn/Dunmurry road around the turn of the 20th century would have seen some fine black matrons grazing the pastures around Killeaton House. These included 'Gipsy Lass of Ruthven', 'Lucy of Portlethen', 'Mayflower of Mulben' and 'Miss Douglas of Culsh'. Some of the early bulls were 'The Squire of Killeaton' and 'The Primate of Killeaton'.

On 24 March 1898 Mrs Grubb, having paid a fee of five shillings, entered a young bull called 'Prefect of Killeaton' at Messrs Robson's great annual show and sale of 'Blackskin' cattle. One hundred and thirty-one animals were catalogued and the high number of entries reflected the growing interest in the Aberdeen Angus breed across Ireland.

Mrs Grubb's 'Prefect of Killeaton', which took the first prize and Silver Cup, was described in a show report as being a "beautiful bull". He came into the ring as Lot 15 and auctioneer, Mr John Robson, did not close a deal on the animal until bidding had stopped at 28 guineas. This sum, one of the highest of the day, was paid by Mr CJ Johnston of Jordanstown. The second prize bull, owned by Miss HH Coultar, Comber, sold to Mr WJ Doloughan, Dromore, for 26 gns, while other bulls in the class realised sums between 9 ½ and 25 gns.

Of all the pedigree Aberdeen Angus cattle owned by Mrs Cambridge Grubb, the most successful in the show ring was a bought-in Scottish-bred female called 'May Empress'. This animal, born 26 February 1896, was bred by Mr AR Stuart, whose farm at Inverfiddich, Craigellachie, was situated in the heart of picturesque Speyside. Under the ownership of Mrs Grubb, 'May Empress' was one of her team that won the 'Best Group of Aberdeen Angus cattle' at the Royal Dublin Society's last spring show of the 19th century, which took place in April, 1899.

A few weeks later, Mrs Grubb's 'May Empress' won the Aberdeen Angus Female Championship on the cattle lawn at Balmoral. Exhibitors at that 1899 show included Major R Claude Cane, St Wolstans, Co Kildare; Charles J Johnston, Greenisland, Belfast; FL Turtle, The Villa, Aghalee; Robert Anderson, The Park, Dunmurry; Thomas J Byrne, Rossmakea, Dundalk; RD Jameson, Devlin Lodge, Balbriggan; and Edward Coey of Droagh, Larne.

The latter breeder brought out a bull called 'Baron Inca' to this Balmoral show, which had come from Scotland inside his dam. This cow had been purchased by Mr Coey from the 'great triumphant of the Aberdeen Angus breed' Sir George Macpherson (1839–1907) of Ballindalloch Castle, N Scotland. At this 1899 Balmoral 'Baron Inca' won the breed Championship. Three years later he was exported to America, where he died at the age of thirteen.

Although 'Baron Inca' had an unbeaten show career in Ireland, this was not the case when his owner exhibited him at the Highland Show in Edinburgh that same year. Perhaps the bull did not travel as well as he could have, or the competition was just too strong but at this show 'Baron Inca' did not get beyond a 'Highly Commended' award.

'May Empress' was also at that Highland Show of 1899, where she gave a good account. In a strong cow class of eighteen entries, she took the second place. While Mrs Grubb may have been disappointed to get a blue rosette at any Irish show, to win one at the Highland was a tremendous achievement.

Mrs Cambridge Grubb (d.1943) also won prizes at Belfast shows with her Large White Yorkshire pigs and when the well-known 'Killeaton' Aberdeen-Angus herd was dispersed, two pure-bred sows and a boar were included in the auction. This sale took place in Messrs. Robson's Royal Victoria Horse Bazaar, Chichester Street, Belfast on Friday 22 November 1901. The advertising notice is reprinted opposite.

The heyday of the Herefords leading to a Dublin world conference

'Herefords are hardy', 'Herefords are great foragers', 'Herefords mature early', 'Hereford bulls type-mark their calves with a white face'. These are just a few reasons given by the Hereford Herd Book Society, in the 1950s, to explain why their breed was the most popular beef breed in the United Kingdom.

There was, during this decade, a growing interest in the breeding of pedigree Hereford cattle in Northern Ireland as an increasing number of dairy farmers milking Friesian cows saw the value of running a Hereford bull. The resulting 'black whitehead' calves were easy calved and heifers had the potential to become great suckler cows.

Alternatively the hardy black whitehead bullocks could, without much fuss, produce the type of carcase which could be broken down to yield succulent roasts and steaks. During this decade, as the demand for the white-faced terminal sire grew, many of the Province's pedigree livestock breeders switched to the Hereford. As one such person put it, "Why give the people oranges when they want apples"!

In 1950, a few years after starting farming, Major JC Mitchell of Cattogs House, Comber, Co Down, made a decision to form a herd of pedigree Hereford cattle. Having registered the prefix 'Cattogs' with the Hereford Society of 3 Offa Street, Hereford, Major Mitchell's early purchases included four 'Free Town' beasts. Modern day breeders and Hereford historians will recognise the great 'Free Town' prefix, which was founded over one hundred years ago and remains, four generations on, with the Bradstock family.

During the first half of the 20th century, when Argentinean ranchers were keen to buy up good Hereford bulls for improving their native 'scrub' cows, several 'Free Town' bulls were taken across the Atlantic. These included one called 'Free Town Contrite' that had been born on 25th October 1947. Before making this trip, this fine bull was awarded the 'Supreme Championship' and 'Reserve Supreme Championships' at the Three Counties and Royal Shows, respectively. 'Free Town Contrite' was also used, for a time, as a stock bull in the herd of his birth.

The four early Herefords purchased by Major Mitchell of Comber included a son of 'Free Town Contrite' called 'Free Town Envoy'. Regarding the females brought in, 'Free Town Tara' was, perhaps, the most successful; she remained in Major Mitchell's herd well over a decade and gave birth to good stock, most notably her 1959-born bull calf.

Exhibited at the Royal Dublin Bull Sales in October 1960 as a yearling, this animal, named 'Cattogs Punch', was placed ahead of over six hundred entries to win the RDS Championship Silver Medal for Best Hereford, the RDS Silver Medal for 'Best Hereford yearling' and the 'Hereford Society's Champion' prize. A growthy bull with a daily liveweight gain of 3.08 lbs, he took the day's top bid of 900 gns.

At Royal Ulster Agricultural Society's annual bull sale, held at Balmoral that year, pedigree Herefords consigned from Major Mitchell's herd gave a good account. At this event Reserve Champion 'Cattogs Pirate' was knocked down, after an enthusiastic bidding battle, to the Marchioness of Dufferin and Ava, Clandeboye, County Down, at 310 gns.

During that year of 1960 the 3rd World Hereford Conference took place in the United States of America. This event, staged every four years, was of global importance and as such, when a decision was taken to hold the 4th Conference in Dublin, the significance of this recognition was not lost on the Irish Hereford Breeders' Association members.

When, after months of careful planning, the 4th Conference week of events got underway, the Northern Ireland breeders were represented by the aforementioned Major Mitchell, in his

'Free Town Contrite' (born 25 October 1947) was the sire of JC Mitchell's Hereford bull 'Free Town Envoy'.

MARBLED BEEF

Of all our breeds of cattle that in which the fat and the lean are most evenly intermixed is the Hereford, and it is for this reason that the picturesque white faces which have their homes in the English Midlands always find such favour with Butchers.

Hereford meat, in the technique of the trade, is always "beautifully marbled" or, in other words, its lean and its fat are very evenly blended, and this renders their joints much more saleable than those of other breeds in which the lean and the fat are not so well mixed.

Quotation from The Farmers' Gazette, August 1901.

role as Chairman and Mr RJ Beck as Vice-Chairman. Two other NI committee members were also present and these were Mr and Mrs Sam Douglas, owners of the Dungannon-based 'Culnagrew' Herd who, at the time, were serving as Secretary and Treasurer, respectively.

Around one hundred and thirty cattle were put on display for the foreign delegates and it was, up to that time, the best collection of Herefords ever witnessed in Ireland. For the northern representatives, it was a matter of satisfaction when one of the Douglas' heifers was awarded a first place and plaque presented by the Argentine Hereford Breeders. This animal's name was 'Culnagrew Primrose'.

Not only did the visitors attend the Irish Derby at the Curragh during that 4th Conference Week, they also travelled north through the beautiful Mountains of Mourne. Having greatly enjoyed the scenery, the forty delegates were ready to see some more good Herefords and accordingly made their way to the aforementioned fifty acre farm of Major Mitchell at Cattogs House.

Having received a warm welcome, the visitors were shown round the farm. Passing under an archway adorned with the image of a Hereford beast, they entered the farmyard where the bull boxes were duly reported as being 'spotlessly clean'. When taken to the pastureland around Cattogs House, the visitors would have seen around eighteen breeding cows running with their offspring and senior stock bull.

When the time came for the delegates to take their leave of Comber and indeed, Ireland, following a successful 1964 Hereford World Conference, they would have done so with a sense of pride in their breed. For those from the north and south who had been involved with the event, it was a memorable week and it took place on the verge of a 'continental invasion'.

Despite the challenges we can, with hindsight, say that those Hereford herd-owners who stayed on course would see their breed, come into favour once more. 'Herefords are hardy', 'Herefords are great foragers', 'Herefords mature early', 'Hereford bulls type-mark their calves with a white face' and finally, 'Herefords have stood the test of time'.

Longhorns at the heart of breeding revolution in 1700s

During the 1840s some fine Leicester Longhorn cattle would have been seen grazing the pastureland of the Seaforde Estate, near Downpatrick.

The cream of those cattle paraded at our provincial agricultural shows during the past 2008 season would not have looked out of place had they been turned out at any such event staged across the globe. Our cattle have gained a worldwide reputation and this has been reflected in the fact that, over recent decades, many Northern Ireland cattle breeders have been asked to judge dairy and beef classes at both National and International shows. The excellent reputation of Northern Ireland's livestock and livestock breeders is indeed, something to celebrate.

That's all very grand but some *Farm Week* readers may be wondering if we always enjoyed this high standing. What about our cattle breeding ancestors way back in the 1830s – were their animals held in such high esteem? Well, according to an essay published in the Quarterly Journal of Agriculture, published in 1836, the cattle in the north of Ireland were a "shapeless inferior sort"!

The writer did, however, exclude those animals belonging to the Longhorn breed from his viewpoint, claiming that they were both "large" and "valuable". Warming to the subject, they went on to describe these cattle as having "straight, level backs, strong bone, plenty of hair, placid eyes, and great substance with thick shoulders, short coupling and backs well covered with beef". Regarding colour, the 1836 commentator in the Quarterly Journal, stated these Longhorns were "mostly red and white with white along the back".

Whereas from the dark ages domestic livestock had procreated under a 'survival of the fittest' regime, during the 18th century the opportunities offered by artificial selection were becoming increasingly apparent. Embracing the principle of 'Like begets like', the old haphazard system was left aside and farm livestock matings increasingly took place on a planned basis.

One of the best-known breeders of Longhorn cattle was Robert Bakewell (1725–1795) of Dishley, Leicestershire, a man described as being "the great originator of the system of improvement of farm livestock by in-and-in-breeding". Although some thought this practice of mating closely related animals an outrage, the enlightened Mr Bakewell understood that it was merely a technique that, wisely used, could help 'fix' the type. When giving advice to his fellow Longhorn Leicester cattle breeders he often said, "Breed the best to the best".

A cattle breeder by the name of Robert Fowler from Little Rollright in Oxfordshire took some of Bakewell's cattle and continued to inbred them to what some believed a 'dangerous extent'. One of Fowler's best-known Longhorn Leicester bulls was called 'Shakespeare', an animal described by William Marshall in the late 18th century as being a "striking specimen".

In 1807 an imported grandson of Mr Fowler's Old Shakespeare was standing at Haypark, Belfast (off the Ormeau Road) on the farm of Lord Donegall (1769–1843). According to a newspaper advertisement the bull could be let at half a guinea a cow and "offered farmers and gentlemen the best opportunity to tap into the best blood that England could boast of".

Later in the 19th century and just before the those valuable Improved Shorthorn cattle from County Durham swept across the north of Ireland, herds of Longhorn cattle could have been seen at Bangor Castle, Castlewellan, Waringsfield, Waringstown and Orangefield. Perhaps the Longhorn 'flagship' herd in the North of Ireland during the 1840s was that owned by Rev William Brownlow Forde. His

cattle would have been seen grazing the pastures around Seaforde House, near Downpatrick.

Having succeeded to the estate following the death of his brother, Rev William B Forde (1786–1856) combined his duties as Rector of Annahilt Parish Church with his responsibilities as estate owner. He was well-known in the farming community and served as Chairman of the Seaforde and Hollymount Agricultural Society. When that body staged its annual show in August 1842 a report stated that Rev Forde was awarded the society's medal for 'Best cultivated crop of drilled turnips'. It also stated that his cattle had been greatly admired, singling out a Longhorn bull and eight fine milch cows.

The following year, on Wednesday 30 and Thursday 31 August 1843 the prestigious Royal Agricultural Improvement of Ireland held its Great National Cattle Show on a piece of ground to the rear of May's Market in East Belfast. Cattle exhibitors were given the choice of bringing their beasts into the showyard on the Wednesday evening or between the hours of five and seven on the morning of opening day!

If an active young Leicester Longhorn bull was seen being led through the gates during these hours, it may well have been that one owned by Rev Forde. On the day of the show, this animal was selected by the judges for the first premium in its class and prize of ten sovereigns. Perhaps, some of the gentlemen or farmers seeing Rev Forde's prize bull at this 1843 show would have been keen to procure one of his sons. Had this been the case, an opportunity presented itself the following year when four sons were advertised 'for sale'. They were out of well-bred dams that Rev Forde had selected from the herds of 7th Baron Farnham from Farnham near Cavan and Mr G Lucas Nugent of Castlerickard in County Kildare.

Looking over old records, it can be gleaned

that prize Leicester Longhorn cattle grazed the pastures of the Seaforde Estate over three decades and as such that location stands out on the breed's heritage in the North of Ireland. Given the success of his cattle, not only in Belfast but in Dublin as well, Reverend William Brownlow was one of our best-known breeders of Leicester Longhorn cattle.

Described as being "guileless and humble, gentle and charitable", it was a cause of great sadness when, on Tuesday 11 March 1856 Reverend Forde passed away. According to an obituary all who knew him could bear testimony to his benevolent and honourable character, adding that he was a beloved father, husband and friend.

The dual-purpose breed that faded from scene in 1960s

When one of Northern Ireland's pedigree Friesian herd-owners happened to remark that the breed was dual purpose, that is for both milk and beef, it brought something of a rebuke from the Society's secretary Mr Bursby who, in an attempt to set the record straight, said it was not a dual-purpose breed but rather "a single purpose breed with a dual result". *Farm Week* readers may be interested to learn that as young man, Mr Bursby (1917–1977) commenced his career in the legal profession.

Mr Bursby's sentiments would have been fully shared by an agricultural writer who, several decades ago, explained the difference between dairy breeds, beef breeds and dual-purpose breeds. Breeders of the first type had gone down the road of improving milk yields, while breeders of the second had gone down the road of producing more meat. Regarding those breeders of dual-purpose breeds, they had tried to go down two roads at the same time and as a result got nowhere!

While Mr Bursby, as secretary of the British Friesian Cattle Society, may have belonged to the 'Dual-purpose, No purpose' school, the same cannot have been said of the Mr AC Burton, Secretary of the Red Poll Society. According to breed promotional material in the Farmers' Yearbook for 1954/1955 "ever since the Red Poll Society was founded in 1888 members of the organisation have always exhorted to place their faith in the breeding of dual-purpose".

After providing a little information on the Red Poll Society and some historical snippets of that breed's history in Northern Ireland, it may be interesting to focus on one particular County Down herd that, in terms of breed activity during the 1950s, was travelling a 'lonely' path.

Basically, the Red Poll breed of cattle originated in East Anglia and

A group of red poll steers at grass.

was the result of bringing two breeds into a common type during the 1840s. These types were the old horned 'Norfolk Red' and the polled 'Suffolk Dun', with the former having gained an excellent reputation as a grazer and the latter as a pail-filler.

What the name Charles Colling is to the Shorthorn breed history, Francis Quartly is to the Devon, Benjamin Tomkins is to the Hereford and Hugh Watson is to the Aberdeen Angus, the name Henry F Euren stands out in the annals of the Red Poll breed. It was largely through his efforts that a Herd Book for these cattle appeared in 1874, with a second being issued in 1877.

The Red Polled Society of Great Britain and Ireland was formed in 1888. Before the arrival of the 20th century the breed's council stated that the Red Poll "was a dual-purpose cow that must be able to give a good account of herself in the dairy, and be able to produce calves which, when steered, will grow well and fatten into fine butchers' beasts".

One hundred years ago, a visitor to Mr Knox Gilliland's farm at Londonderry would have been rewarded with the sight of five such cows, purchased in April 1905 from Mr Garrett Taylor's renowned herd in Norwich. These cows, 'Damsel, Lema', 'Curls 2nd', 'Ring 2nd' and 'Dulcie' had been purchased for

15½ gns, 15½ gns, 17 gns, 17½ gns and 18 gns, respectively.

After the 1914–1918 war some more highly-bred Red Polls would have been seen on the farm of Mr Eric C Lindsay of Keady. During 1924 several of his animals came from the best-known prefixes in the Red Poll herd book such as 'Framlington' (William Woodbridge, Framlington), 'Marham' (Thomas Brown and Son, Marham Hall, Norfolk), and 'Kirton' (Stuart Paul, Kirton, Ipswich). This latter herd was established in 1899 and numbers increased to around 500 head. 'Kirton Lilsome' was the name of Mr Lindsay's cow out of this herd and she was successfully paraded at the Royal Ulster Agricultural Society's Balmoral show in 1924.

Although the pleasing spectacle of Red Poll cattle being led by white-coated handlers across the Balmoral turf disappeared during the 30s and 40s they were back, albeit in low numbers, during the 50s. In this decade the two Ulster breeders who carried the breed's banner at our leading show were Major-General F Beaumont-Nesbitt, Lisnabrague Lodge, Scarva and Major WS Brownlow, Ballywhite, Portaferry.

When the former's attested Red Poll herd at Lisnabrague Lodge, comprising fifteen pedigree cows, two young bulls (licensed), four pedigree in-calf heifers, and twelve pedigree maiden heifers, was dispersed by South Down Auctioneers Ltd, Newry, on 2nd March 1956, it left Major William S Brownlow somewhat on his own. Before mentioning some of the Major's RUAS prize-winners, it may be useful to provide a little information on their owner.

William Brownlow was born in Winchester during 1921 and as a young man came to live in the family's Ballywhite House, situated on the side of Strangford Lough. Having completed his education at Eton College,

William Brownlow joined the army and served during the Second World War. When his father became unwell, William left the army as a Major in December 1953 and returned home to oversee work on the family estate in County Down.

In the years that followed 'The Major' was to serve as Lord Lieutenant for the county and played an active role in rural life. He had a great love of field sports and would have been a familiar face at various point-to-point meetings (astride the grey mount 'Chipolata') or watching his racehorses compete at the Down Royal and Downpatrick courses.

Regarding his Red Poll cattle, Major Brownlow's first successes at the Balmoral Show came in 1956 when he exhibited the cows 'Westwoodhay Sweep 6th' and 'Westwoodhay Bertha 2nd'. These were all bought-in, foundation animals, born and bred in the herd of Mr John Henderson, Westwoodhay. Other pedigree Red Polls purchased for the Major's 'Ballywhite' herd came from the 'Seal Point' and 'Aughton' prefixes.

Major William Stephen Brownlow (1921–1998) of Ballywhite House, Portaferry, continued breeding pedigree Red Polls into the mid sixties. The 'Ballywhite' herd, comprising 15 cows, 11 heifers and 1 stock bull, was dispersed by auctioneers Osborne, King and Megran, Castle Street, Lisburn on Thursday 21 October 1965.

Details on Major Brownlow's Red Poll herd dispersal held during October 1965

DISPERSAL SALE BY AUCTION

Pedigree Red Poll & Crossbred Cows, Heifers, Suckling Calves, Young Stock, Milking Parlour, Machinery and Baled Hay.

AT DERRYHILL, BALLYDUGAN, DOWNPATRICK, CO DOWN ON THURSDAY 21 OCTOBER 1965 AT 11 AM.

On the instructions of Major WS Brownlow

Pedigree Red Poll — 15 cows and Heifers.

Cross Bred Cows — 18 cows and Heifers, some with calves at foot.

Young Stock — 64 Bulls and Heifers from 3 months to yearlings.

Bulls — 2 Hereford and 1 Red Poll.

Hay — 1,000 Bales Hay.

Machines — 1954 Fordson Major Tractor, Pheasant Tay Ted-der; Buckrake; Trailer and Roller Mill.

Milking Parlour — Gascoigne 2-unit milking machine; Gas-Coigne Rotorfreeze cooler; Sterilizing Chest; Wash-up Trough; 12 milk Cans.

NB. The milking parlour and Dairy equipment is situated at Ballywhite, Portaferry.
ON VIEW Wednesday 20 October from 10am till 5pm and morning of sale.
A detailed catalogue price 1/- may be obtained at the office of the auctioneers.

A mobile canteen will be in attendance on sale day.

"other methods of slaughter were just as much entitled to be described as being 'humane' as the mechanical killer."

Poleaxe advocates refuse to lie down in long-running debate

Visiting Belfast's Stewart Street Abattoir during its opening year of 1912, one would have seen animals being dispatched by poleaxe. This was not a sight for the faint-hearted; the beast, having had its head well secured to a metal ring, was delivered a deft blow to the head. Rendered senseless, it was then ready for sticking.

Although this method of slaughter was acceptable for many centuries, the advent of the 'Humane Killer', around the turn of the 20th century, brought the practice of poleaxing into question. With the new technique, an exploding cartridge would drive a captive bolt of hardened metal into the skull, felling the animal instantaneously. The humane killer could, its exponents claimed, be used with precision, unlike the wieldy poleaxe.

When the province's butchers were not quick to embrace the new instrument, they were labelled 'cruel men' in letters to the press. Not only that but the claim was also made, during January 1913, that 'raw lads' and 'casual persons' were being used to slaughter animals in the factory. Faced with such criticism, the Executive of the National Federation of Meat Traders Association presented their case.

Taking the 'Raw Lad' allegation head on, they claimed it untrue and stated that slaughter men had years of practice before they became "first hands" or "foremen". These workers had to "become efficient in the use of both chopper and axe on the carcase before they were allowed to touch the live animals", they added.

Addressing the claim of poleaxe inaccuracy the Federation of Meat Traders stated that "the vulnerable portion of a bullock's skull was of such proportion that perhaps the wonder was that one could make a mistake". Regarding the use of the gun, they stated that cartridges had been known to misfire and sometimes two or three attempts had been needed to render a beast unconscious.

The Humane Killer versus Poleaxe debate continued into the 1920s, with the Ulster Society for the Prevention of Cruelty to animals favouring the former and the Master Butchers and Journeyman Butchers the latter. A meeting of all three groups took place in the Wellington Hall, Belfast on Friday 9 January 1920.

The gathering was held under the auspices of the USPCA, to protest at the method being used at Belfast's Stewart Street Abattoir and to urge the substitution of more humane reforms, by the introduction of local government bylaws. Speaking in favour of change, Mr RJ Lynn MP stated the old-fashioned methods of killing animals should be abolished and in his judgement it was a disgrace that they were being used in an enlightened and progressive city like Belfast.

Mr Lynn also stated that people had been told the 1914–1918 war had created a brutal spirit among the nations but he thought, given the large attendance at the meeting, this view had been repudiated. By so many being gathered to discuss matters affecting the humane treatment of animals it proved that the heart of Belfast was, in fact, very human.

The fact that the representatives from the meat industry did not want to endorse the humane killer did not, of course, mean that they didn't care about animal welfare. They had other concerns. Although the view that using the Humane Killer was as easy as using an umbrella, Mr J Mairs, speaking on behalf of the butchers, stated that this was not the case as every animal's skull was not the same.

Mr Mairs told those gathered that he had used the humane killer more than anyone in the hall and it would not knock down a big bull. "The the only thing it knocked down", he added, "Was what any man could have floored with his cap!"

Another problem with the Humane Killer, expressed by the journeymen butchers at that 1920 meeting in Belfast's Wellington Hall, concerned meat keeping quality. It was stated that, according to a number of eminent scientists in Belfast, the shelf life of meat from animals killed by the humane killer was not as good as that when the old method was employed.

It was estimated that meat produced by the new method would not keep longer than 36 hours, because animal killed in this way did not bleed as well. With the modern method of slaughter it was believed more blood remained in the tissues, thus bringing forward the onset of decomposition. If only on the grounds of providing good meat for the citizens of Belfast, the butchers would reject the humane killer.

This Wellington Hall meeting failed to bring about agreement and as such the debates for and against the Humane Killer in the Stewart Street Abattoir continued over the next few years occupying many Belfast Corporation councillors. Those on the Markets Committee of this body had responsibility for running the abattoir. When a proposal to introduce the humane killer into the Stewart Street premises was moved and seconded at a particular meeting in June 1923, it caused something of a stir.

Perhaps there were some members who thought the decision would be 'rubber stamped' by passing the minutes at the next meeting, but this was not to be the case. Some councillors were still vehemently opposed to the humane killer's introduction and the meeting attracted protests by representatives of the Belfast Master Butchers' Association, Ulster Bacon Curers' and Belfast Journeyman Butchers' Association.

Addressing the markets committee, Mr Hugh Bowman of the MBA, stated that if the killer was introduced, the butchers were considering erecting an abattoir of their own, adding that they had a site. Falling in behind, Mr S McKitterick of the Journeyman Butchers, said his members were 'not going to stand up and be shot at'. Regarding increased costs if the killer was introduced, Mr McKitterick stated that

ammunition for it would cost around £500 per annum. The Journeyman Butchers' took a strong line regarding the matter and actually passed a resolution forbidding its members from using the humane killer in the Belfast plant.

While the Meat Industry representative spoke with one voice, the same cannot be said of the Corporation's market committee members who had a lot to say about the proposal. Councillor Magowan stated that it was a big mistake for the committee to interfere with the abattoir and put forward an amendment that the decision taken at the previous meeting to introduce the humane killer be deleted from the minutes.

Expressing his opinion that the pole-axe was not only an antiquated implement but caused cruelty, Councillor Wilson maintained that the British House of Commons was supposed to compel butchers to use the mechanically-operated instrument and the great majority of the people of Belfast looked to the Corporation to have this reform brought into place.

Alderman Mrs McMordie MP took, one

could say, a more balanced view stating that she had recently visited the public abattoir for a demonstration. Three methods of slaughter had been undertaken and she had not noticed any difference between them, except in the case of pigs, when the humane killer had not proved effective. Mrs McMordie also expressed the view that the other methods of slaughter were just as much entitled to be described as being 'humane' as the mechanical killer.

At that 1923 Markets Committee meeting, Councillor Kelly moved that the matter be adjourned for a month, so that the new members of the Corporation might have the opportunity of visiting the public abattoir to see the various methods of killing cattle. This was passed and so the long running poleaxe versus humane killer debate was saved for another time.

Days, weeks, months and years passed before real progress was made and as so often happens in these matters, one could say that common sense prevailed. The meat traders did not desert the Belfast Meat plant and the Journeymen Butchers rescinded the resolution previously taken, forbidding its members from making use of the mechanical killers in the slaughter of cattle in the Stewart Street Abattoir.

The time came when butchers acknowledged the advances made in the perfecting of the humane killer, with regard to reduction of misfires and increased speed of reloading. By 1927 the Journeymen Butchers' Association, who met in East Bridge Street, Belfast, were of the opinion that the humane killer was essentially a 'mechanically operated pole-axe'.

As the 1930s approached there was less talk of 'raw lads' 'cruel men' and 'stone age methods'. The growing acceptance of the Humane Killer would in time ensure that the poleaxe, once held in such high esteem by the butchers in Belfast's Stewart Street Abattoir, would be consigned to a place in history.

Not a bull in the china shop but almost a cow in the scullery!

Having been instructed to drive the family cow to a neighbour's bull, two young brothers set out from their farm on the Moira side of Magheralin early one morning back in the 1940s. Everything was going swimmingly until they approached a row of terrace houses, one of which had the front door lying open. To the horror of the hapless drovers the cow suddenly bolted into the house and managed to jam herself in the narrow passage between the wall and wooden staircase.

After what seemed like an eternity to the brothers, the cow reversed out of the dwelling to the sound of splintering timbers. Had anyone in that row of terrace houses looked out their bedroom windows on that particular morning, they would have seen two boys and a 'bulling' cow galloping off in the direction of Lurgan.

Well, that was over sixty years ago and now is a good time to bring everything out into the open. The house 'visited' by the cow was owned by a gentleman called William Henry Irwin. Perhaps, down through the years the Irwin family have often wondered what happened to the staircase of their house back in the forties. Today in *Farm Week* the truth has been revealed.

On that particular occasion a whole lot of

trouble could have been avoided had the farmer, instead of relying on his sons, simply called in the AI man or rather, 'the bull with the bowler hat'. By the late 1940s, the Northern Ireland Ministry of Agriculture had gathered a stud of bulls at Desertcreat, near Cookstown. Although these animals where all described as 'Dairy Shorthorns' one of them, a dark red, had a dash of Lincolnshire Red Shorthorn blood running through his veins.

Following the emergence of the Shorthorn breed in County Durham early in the 19th century, some red bulls went south to the Lincolnshire area. Some of the early breeders kept their own private 'pedigree' records before coming together, in 1895, to form the Lincolnshire Red Shorthorn Association. The new name did not go down well with some members of the Shorthorn Society of the United Kingdom of Great Britain and Ireland (est. 1874) who, at that time, were recording the pedigrees of their cattle in the *Coates's Herd Book*. They argued that the new association had no right to use the word 'Shorthorn' in their title but it was to no avail; The Lincolnshire Red Shorthorn Association published their first herd book containing 203 bulls and 917 females in 1896.

In July 1935, following an agreement between

the Shorthorn Society and the Lincolnshire Red Shorthorn Association, the former body took on responsibility of registering Lincolnshire Red Shorthorn Cattle in a separate section within the *Coates's Herd Book*. The pedigrees of all of the Lincolnshire cattle were to be denoted by the letter 'L' in front of their names or numbers. The first Lincolnshire Red Shorthorn beast registered in *Coates's Herd Book* under the new system was 'L1 Wolferton Ruby 18th' (c.8 November 1935) owned by His Majesty the King, Sandringham, Norfolk.

Eight years later it was a case of 'all change again' when, following a meeting held on 21 January 1943, there was a parting of the ways between the Shorthorn and the Lincolnshire Red Shorthorn breeds. In that year's *Coates's Herd Book* those breeders of Lincolnshire Red Shorthorn cattle who wanted to continue registering their cattle, were told to drop the words 'Lincolnshire Red' and renounce all rights to exhibit cattle in classes for Lincolnshire Red Shorthorn cattle at agricultural shows. It was further laid down that any cattle entered in *Coates's Herd Book* that had Lincolnshire Red Shorthorn blood, would have a mark or a letter beside their pedigree name.

In the early 1940s, Cambridge University ran

This fine Lincolnshire Red Shorthorn cow was photographed in the late 1940s.

a dairy herd on their farm at Huntingdon Road, Cambridge. It was rich in Lincolnshire Dairy Shorthorn blood and as such the pedigrees of their twenty-four calves (13 heifers and 11 bulls) all carried the stipulated designation. One of the male calves was called 'Cantab Special Blend', born 26 September 1943 and this was the previously mentioned bull standing in the Northern Ireland Ministry of Agriculture's Artificial Insemination Centre at Desertcreate, Cookstown, in the late 1940s.

One of the Ulster farmers who incorporated a touch of Lincolnshire Red Shorthorn blood into their herds during these years was Mr William Houston of Ballycloughan, Broughshane, County Antrim. When the 97th volume of *Coates's Herd Book* came off the press it contained the pedigree of one of Mr Houston's heifer calves. According to this old book, this animal, called 'Quarrytown Winne', was born on 7 January 1950 and was a daughter of the AI bull, 'Cantab Special Blend'.

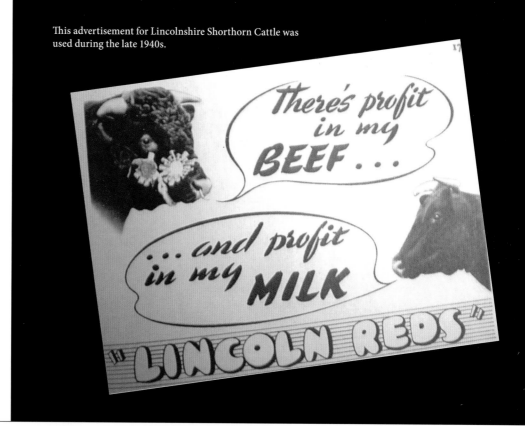

This advertisement for Lincolnshire Shorthorn Cattle was used during the late 1940s.

MILKING DETAILS ON 'CANTAB MAUD 29TH', DAM OF DESERTCREATE AI BULL 'CANTAB SPECIAL BLEND' (BORN 26 SEPTEMBER 1943)					
Lactation	Date of calving	lbs	Kgs	Days	% butterfat
1st	3.9.1942	7414	3365	314	3.51
2nd	26.9.1943	8066	3659	311	3.93
3rd	29.9.1944	7542	3421	292	3.46
4th	28.9.1945	9452	4287	315	
5th	1.4.1947	10255	4651		3.75

"the milk was sold exactly as it came from the cows and nothing had been added to it"

Before the magistrates for low butterfat count

This non-pedigree Shorthorn cow was photographed during the 1920s.

It was before Keady justices in October 1927 that a certain County Armagh farmer was charged with selling milk that was deficient in butterfat by just 0.33 percent. Samples taken from milk off his farm gave a result of 2.67 percent, slightly under the stipulated minimum of 3 percent.

Having been sworn in the defendant admitted that there was a moment when the milk was not under his personal supervision, but he was adamant that it "had not been tampered with in any way" and that, "the milk was sold exactly as it came from the cows and nothing had been added to it". He further stated that all he could do was to test his cows separately and when he found which cow was giving weak milk, he would get rid of her.

At the hearing, Mr W Gore Moriarty RM, the Chairman, said that the magistrates quite thought the defendant did his best to carry out his business in a thoroughly honest way. They would give effect to his high character and past good

Name of cow	Colour	Year of birth	Owner	Yield lbs	% b/fat	weeks
Bluebell	Blue	1919	John Allen, Magheryarville, Milford	8050	4.66	34
Jenny	Red and white	1915	James Bothwell, Killylea	7679	3.73	37
Buttercup	Red and white	1920	James Carter, Derryall, Portadown	6485	3.93	45
Violet	Red and white	1918	Thomas George, Salters Grange	6594	3.81	33
Rose	Red, little white	1915	Miss Rebecca Gray, Ivy House, Tynan	12750	3.41	44
Nellie	Red, little white	1917	David A Greer, Drumatee, Markethill	7500	3.57	33
Sweet Briar	Red	1918	Robert Hamilton, Crossdall, Middletown	8190	4.09	45
Nellie	Red and white	1915	Thomas H Hardy, Woodland, Richill	7231	3.65	32
Loughgall Queen	Red	1916	SP Hendren, Ardress, Annaghmore	7595	4.30	35
Espilon	Red	1914	Captain V Irwin, Tynan	6296	4.03	35
Mollie	Red and white	1919	George Leeman, Grangemore	10458	4.04	36
Primrose	Roan	1918	Pooler Leeman, Rose Cottage, Killylea	6153	4.33	38
Molly	Blue	1917	Thomas E Maginnis, Breagh, Portadown	6135	3.67	30
Cowslip	Roan	1914	Mrs AM Mitchell, Killycopple, Collone	7371	3.9	30
Daisy	Roan	1918	Mr H Rollston, Polnagh, Killylea	7045	4.31	32

record. There had, however, been a breach of the law and he would be fined only 2s 6d with 10s 6d charge for the analyst's certificate and court costs.

"When fair occasion call, 'tis fatal to delay" would have been a most apt quotation for the secretary of Armagh Milk Recording Association, had they been reading about the aforementioned court case in their daily newspaper. Seizing the opportunity, they may well have visited the farm in question. Perhaps, the conversation could have gone as follows:

Secretary: " Oh Hello Mr _ , I hear you had a bit of trouble with your milk recently".

Farmer: "Yes".

Secretary: " Yes, er … indeed – that was most unfortunate – you had to part with over twelve shillings – dreadful business altogether, especially as there was no tampering on your part".

Farmer: "Yes".

Secretary: "Perhaps, I may be of assistance to you".

Farmer: "Yes?"

Secretary: "Absolutely … I belong to the Armagh Milk Recording Association, and we have the answer to your butterfat problems, all you need to do is …"

Addressing the conundrum of butterfat variation, author, GA Garrad, NDA (*The Principles of Dairy Farming*, published 1926), stated that herd-owners sometimes received complaints about the BF percentage in their morning milk, especially during the months of May and June when there was abundant grass.

To counteract this, he suggested making the evening milking later, or the morning milking earlier, so that the intervals between them were less uneven. In other circumstances, Mr Garrad added, the complaints regarding butterfat, were "due to a large number of newly calved cows having recently come into the milking shed".

But supposing there continues to be a problem with the milk and it is not to do with morning milk, early grass or a large number of fresh calvers, what would Mr Garrad have to say about this? Well, given this scenario, he suggested introducing some Channel Island cows into the herd at a ratio of one Jersey (or Guernsey) to every eight or nine other cows. "The quality of milk", the author asserted, "is a matter of breed and individuality of the cow", adding "it was desirable to have the milk of each cow in the milking shed sampled and analysed; in this way cows that systematically produce low-quality milk can be detected."

So, for the progressive dairy herd owners in the 1920s, milk recording was the way to go and, as usual, a growing number of farmers from Northern Ireland were keen to start. When the Northern Ireland Ministry of Agriculture published volume one of its register of Non-Pedigree Dairy Cattle, which contained particulars on cattle registered up to 31 December 1925, over one thousand of the Province's farmers were involved with the work being administered through forty-one milk recording associations.

Following his interaction with Keady magistrates, the previously mentioned farmer may well have been persuaded to join the Armagh Milk Recording Association. Details on some of its members' non-pedigree milk recorded cows are printed in the table opposite.

From the days when Roscommon sheep were present even in Russia

This Roscommon Ram the property of Mathew Flanagan of Tomona, Tulsk, Co Roscommon.

From a county of rolling hills, peat bogs and bleak, unsheltered farms, the Roscommon breed of sheep emerged back in the 'mists of time'. Not pushed either for the show-ring or butcher, this variety was gradually and naturally improved under direction of experienced flock masters. They applied that old saying among farmers of the West, "Select good sires and the flock will follow".

In time the Roscommon breed was held in high esteem for its ability to produce a fleece of long silky wool and carcase of exquisite mutton. Writing many years ago in a book called *Livestock of the Farm*, Professor James Wilson MA, BSc, stated, "The sand land of Roscommon with the intelligent judgement of an industrious people has now produced as sheep of excellent class".

Although the breed was well known as the 'Roscommon' by the 1870s, it was not until 1895 that a society was formed. The following year the first volume of the *Roscommon Long-Wool Sheep Flock Book* was published. By coming together to register their stock, these members were able to fix the type even further.

One hundred years ago a prime Roscommon sheep would have been described as follows, "The head is hornless, and gaily carried; the face long and white, with or without a tuft of wool on the forehead; the muzzle of the ram strong; the ears fine in texture, white or pinky, and of medium length; the tail broad, and well set in; the fleece white, long and heavy, with a broad staple and bright texture."

Today, when we visit the Province's agricultural shows, there are opportunities to see sheep belonging to different breeds. Each variety has its own unique history going back many years to the time when, under the influence of certain people, a foundation was laid. The author, Beatrice Potter, did sterling work with the Keswick

These Roscommon ewes were photographed around 100 years ago.

breed in England's Lake District while the English Leicester can boast of its association with the famous livestock improver Robert Bakewell of Dishley, Leicestershire.

Henry Dudding, who dispersed his flock in 1913, has been described as the greatest of the 'modern' Lincoln breeders; whereas the name 'Alfred Mansell' is synonymous with Shropshire breed history. At this point some *Farm Week* readers may be asking if there were any 'big names' associated with the Roscommon breed. Well, when perusing old newspaper reports and text books the name Matthew Flanagan of Tomona, Tulsk, County Roscommon, stands out. In addition to running his own flock, Mr Flanagan served as secretary of the Roscommon Sheep Breeders' Association.

Published in 1909, the book *Stephens' Book of the Farm* provides some information as to the management of Matthew Flanagan's flock of Roscommon sheep. It records that his 'wedder hoggets' were sold for killing in November and December at eighteen months of age, yielding carcases from 27 lbs to 32 lbs per quarter.

Around one hundred years ago the Roscommon ewes on Matthew Flanagan's holding were given about 1lb each per day of a mixture of cake and oats for a short time

before lambing. After parturition they received about 2 lbs of linseed cake and crushed oats, again only for a short period. The remainder of the year the ewes received no hand feeding, relying only on their ability to forage. Roscommon ram lambs being sold from Matthew Flanagan's flock at Tulsk, were, according to the aforementioned book, taught to eat cake at an early age. They had a small allowance each day up to eighteen months of age, at which time they were sold locally from £7 to £12 each. Some of Mr Flanagan's Roscommon sheep were exported to Russia with excellent results. It was reported that the manager of the Moscow Society Estate appreciated their "hardiness, early maturity and excellent rustling propensities".

Having established some of Matthew Flanagan's credentials as an historical figure in the Roscommon breed's history, *Farm Week* readers may now be asking, were there any 'big names' from the north of Ireland associated with the breeding of Roscommon sheep, a century ago? The answer to this question would have been 'most definitely'. A gentleman called Mr William Moore of Eglish, Dungannon, kept a flock of Roscommons on his holding and for successive years, brought his sheep to the annual sales of Robson's Mart, Royal Victoria Horse Bazaar, Belfast.

That company's twenty-first seasonal sale kicked off on Friday 1 September 1899 when 700 rams, ewes and lambs were to be offered for public competition. These sheep hailed out of several recognisable flocks and the following

Border Leicester rams 3 gns to 6 gns; ram lambs 30s to 71s; ewes 33s to 38s 6d; ewe lambs 20s to 27s; Shropshire lambs 70s to 6 gns; ram lambs 29s to 63 s; ewes 36s 6d to 40s; ewe lambs 23s to 29s 6d; Roscommon ewes 38s 6d to 51s; cross-bred ewes 28s 6d to 34s 6d; Blackface ewes 20s to 27s; wedders 18s 6d to 27s 6d.

Results from Robson's annual sheep sale held on Friday 1 September 1899 at the Royal Victoria Horse Bazaar, Belfast

FAT SHEEP

Two or three score of Roscommon wethers all four or five years-old and carefully fed on hay, oats and turnips to be sold at Grocer's-Hall, midway between Newtownlimavady and Derry and within one mile of the mail coach road.

Reproduction of advertisement dated 31 March 1812

breeds were represented: Shropshire, Border Leicester, Blackface and Roscommon.

According to a sale advertisement, the consignment of Roscommon sheep were from the aforementioned Mr William Moore's flock at Dungannon, and it comprised '100 prime ewes and hoggets, two and three-years-old'. Entry fees for ram and ram lambs entered to be shown singly were set at one shilling each, while pens of ewes, ewe lambs and ram lambs (not exceeding six head) were set at two shillings per pen.

The sale opened up at 11 o'clock on the day, and Mr John Robson officiated as auctioneer. Prices for all sheep forward, including Mr Moore's batch of Roscommons, are published above.

"it forms too prominent a feature of man's vanity to affect to know much more about horse than he really does"

The skill in buying a plough horse is to know its age

Plough! Plough!! Plough!!! This was the message that went out to Ulster farmers during the First World War, at a time it was feared the nation would be starved into surrender. Local newspapers carried notices exhorting landowners and tenant farmers of Ireland to raise their output and assurances were given that "never again in the history of this country would agriculture be allowed to wane".

Providing something of a financial incentive to farmers, one company, White Tomkins and Courage Ltd of Tandragee, London, Liverpool and Boston, offered generous prizes to those farmers who made the largest increase to their acreage under White Oats in proportion to the size of their entire holdings. The prize money, amounting to £500, was spread over five different groups, the first of which covered holdings of five acres or less, and the last farms that had 151 acres or more.

Responding to this call those farmers, having made the decision to turn over old pastureland, may have considered it an opportune time to invest in some new field implements such as those made by R Hornsby and Sons Ltd, Grantham, Ramsome, Sims and Jefferies Ltd, Ispwich

or J and F Howard, Bedford. Machines from these English manufactures were, at the time, readily available in Ireland.

During the Second World War when, once again, farmers were exhorted to 'Plough, plough, plough,' the tractor was on the scene, but back in the first decade of the 20th century, the reliance was still very much on the farm horse. Before putting their hands to the new plough, some of these wartime speculators may have decided it was high time to go all the way and purchase a new or additional farm horse, either at a public auction or by private sale. Although it was a time for farmers to throw off their hesitant ways, they needed to be careful when choosing their new farm horse.

"Unfortunately", Professor Scott recorded over one hundred years ago, "it forms too prominent a feature of man's vanity to affect to know much more about horse than he really does". The novice farm horse buyer who did not seek professional advice was, the Professor concluded, "like a person who acted as their own lawyer; they had a foolish client"!

Although most members of the horse-dealing profession conduct their business in

It is not possible to tell a horse by looking over a

an upright manner, a few unscrupulous characters have been less than honest in their trade. Just as this is the case today so, we can safely assume, it was back in the first decade of the last century, when our farmers were

Inspecting the teeth helps tell a horse's age.

buying in animals for extra ploughing.

Writing in an old book in the 'Hints to horse-buyers' chapter, the aforementioned Professor Scott drew attention to such characters, stating they were never lost for words. If, for example, one said to such a dealer that his horse had a plain head they would have answered by alluding to the four times British Prime Minister,

"What's that you say? He has a plain head? Well, so did Gladstone and him with the best brain in Europe"!

Again, if the potential buyer had said the horses head was too big, the dealer would have off-set this criticism by saying, "Aye, well sure hasn't he a grand neck for carrying it". If a fault regarding the hind leg action of a gelding was pointed out the dealer would retort, "Oh never mind his hind legs Sir, if he moves the fronts ones well the back ones will follow". As the learned Professor was at pains to point out, "there was no point in arguing with a horse dealer". A runny eye would have been caused by a bit of chaff removed the night before and so on and so forth.

A bit of talk from such a dealer was one thing but there were some who, in attempting to reduce the horse's appearance of aging, resorted to such malpractices as 'Puffing the Glims' and 'Bishoping'. As a horse gets older the cavity above its eye socket gets more pronounced but as some of the old dealers found out, this particular indicator could be temporarily offset by making a small incision in the skin through which a small straw could be inserted. By this means, air could then be blown into the cavity and the socket would become shallower, thus years would be 'rolled back'. Back in the early 20th century knowledgeable horse buyer could detect a case of glim puffing by gently pressing the cavity with a finger.

The old dealers knew that it was unlikely that anyone would buy a horse without first having a look in its mouth and this is where 'Bishoping' came into play. This involved using a 'graver' to cut out the cavity normally seen on the plain surface of the corner teeth of a seven-year-old. The process could then be polished off by applying a hot iron, therefore making the mark more permanent. In the same way that the 'puffing of the glims' could be detected by those who knew to look out for it, so too was 'Bishoping' uncovered by those with a good knowledge of equine dentition.

Back in 1915 a 'poem' was published which eloquently guided those people estimating the horse's age by looking in its mouth. At the time it was recommended that children learn it by heart because then, "they would be able to tell the age of every horse in the district". This illuminating poem is published alongside.

The Horse's Age

To tell the age of any horse,
Inspect the lower jaw, of course,
The six front teeth the tale will tell,
And every fear and doubt dispel.
Two middle 'nippers' you behold
Before the colt is two weeks old.
Before eight weeks two more will come;
Eights months the 'corners' cut the gum.
Two outside grooves will disappear
From middle two in just one year.
In two year from the second pair;
In three, the corners, too, are bare.
At two the middle 'nippers' drop;
At three, the second pair can't stop,
When four-years-old the third pair goes;
At five a full new set he shows.
The deep black spots will pass from view
At six years from the middle two.
The second pair at seven years;
At eight the spot each 'corner' clears.
From middle 'nippers' upper jaw
At nine the black spots will withdraw
The second pair at ten are white;
Eleven finds the 'corners' light.
As time goes on the horsemen know
The oval teeth three-sided grow;
They longer get, project before,
Till twenty, when we know no more.

Aristocracy with a sublime talent for breeding splendid Berkshires

"although swine were vulgar animals they did have highly aristocratic patrons"

This Berkshire boar was photographed more than 80 years ago.

Having been described as a "master of English prose" the author PG Wodehouse (1881–1904) has brought great enjoyment to readers for more than seventy decades. Several of his stories feature the hapless Bertie Wooster and his wise valet, Jeeves, while the main character in several other titles is the "monocled dandy and socialist" known by the surname 'Psmith' – pronounced with a silent 'P'.

Another much loved fictional character in several books by PG Wodehouse is Clarence 9th Earl of Elmsworth, whose seat was Blandings Castle set in rural Shropshire. Largely oblivious to many of the comic antics going on around him at Blanding Estate, most of the Earl of Elmsworth's attention was taken up with his prize–winning pedigree Berkshire sow, named 'Empress of Blandings'.

Often Clarence 9th Earl of Elmsworth was to be found fondly gazing over the sty door as the 'Empress' tucked into her hash of bran, acorns, potatoes, linseed and swill. This sight, according to PG Wodehouse, "made his heart leap in the same way as that of the poet Wordsworth used to do when he beheld a rainbow in the sky".

When not in the vicinity of the 'Empress of Blanding's sty, Lord Elmsworth may well have been in the library, thinking about how her life could be improved in such as way as to ensure many more rosettes when the next agricultural show season got underway. The Earl did, however, have rather an overactive imagination, especially concerning his archenemy Sir Gregory Parslow.

This latter gentleman also owned a Berkshire sow in form of 'Pride of Matchingham' and competition each year between this animal and 'Empress of Blandings' was intense. It was the belief of Lord Elmsworth that Sir Gregory was not above having the 'Empress' stolen, nobbled or just about anything in order to prevent her from winning the 'Best fat pig' class at the annual Shropshire show.

All this was fiction, however, those interested in Ireland's pig history may be interested to know that we did, in actual fact, have an Earl who showed Berkshire pigs each year at the North East Agricultural Society's shows, which took place (prior to a move to Balmoral in 1898)

This Berkshire sow was photographed more than 80 years ago.

in the market area of East Belfast.

This gentleman was Thomas Fortsescue, 1st Baron Clermont (1815–1887) of Ravensdale Park, near Newry. A report on the 1858 Belfast show stated that although swine were vulgar animals they did have highly aristocratic patrons, going on to make special mention of a splendid silver medal winning black Berkshire boar, exhibited by Lord Clermont, which won general admiration.

When the fictional character Lord Elmsworth was at his happiest in the books by PG Wodehouse, he was to be found in the Castle's library with his head buried in the pages of his favourite book entitled *Care of the Pig* by Whiffle. If, during the 1880s, one had asked the real Lord Clermont to recommend a good porcine book, he may have mentioned a then recently published title by James Long.

Published by L Upcott Gill, London it was known as the *Book of the Pig* and dealt with selection, breeding, feeding and management.

In a chapter on the Berkshire breed, which originated in the valley of the Thames, the aforementioned volume stated that the Berkshire had been gradually gaining ground throughout England, Scotland and Ireland and as the exhibition system grew had found admirers in almost every district.

Lord Clermont was one of our leading exhibitors of Berkshire pigs in Belfast for a period spanning four decades and there can be little doubt that his Scottish-born land steward, Mr McClelland, who worked at Ravensdale Park for 42 years (between 1845–1887), also played an important role in all of this.

The esteem in which Lord Clermont's Berkshire pigs were held was reflected in a

1870s show report that he had been offered £75 for three "magnificent" animals by an American speculator. "Lord Clermont's Berkshires", the report continued, "were well known not only over the Province but throughout the country", adding that as far as this breed was concerned "there has been no-one to oust him in the position he has long and deservedly occupied as a breeder of swine in the country".

As is sometimes the case with those who take up the challenge to show pedigree livestock, Lord Clermont knew what it was to experience a disappointment. Take, for example, the Belfast show held in the Municipal Markets, Chichester Street, in 1877 where one of Lord Clermont's exhibits was a "magnificent" boar. This animal was expected to take high honours but died having succumbed to the fatigue of travelling to the show in the intense heat.

This temporary setback did not, however, stop Lord Clermont from bringing pigs to further North East Society's shows held in Belfast, as can be seen the results for the 1880 event, which are printed alongside.

North East Agricultural Association of Ireland
annual show in Belfast of

Live Stock, Poultry, Machinery

and various manufactures
to be held on

Thursday and Friday,
24th and 25th June 1880

Swine
Coloured

Best Boar over twelve-months-old
1. Lord Clermont, Ravensdale Park, Newry
2. Lord Clermont, Ravensdale Park, Newry

Best Boar over six months and not exceeding twelve months
1. Lord Clermont, Ravensdale Park, Newry
2. Lord Clermont, Ravensdale Park, Newry

Best Breeding Sow, in-pig and having had a litter within six months, over eighteen-months-old
Lord Clermont, Ravensdale Park, Newry
GN Callwell, Lismoyne, Dunmurry

Best Breeding Sow not exceeding eighteen-months-old
Lord Clermont, Ravensdale Park, Newry
Lord Clermont, Ravensdale Park, Newry
hc David Glenn, Kilfennan, Waterside, Londonderry

Best pen of three breeding pigs of same litter under ten-months-old
Lord Clermont, Ravensdale Park, Newry
David Glenn, Kilfennan, Waterside, Londonderry

"No man should be allowed to be President, who does not understand hogs"

Paying homage to the Large White Ulster – a magnificent breed of pig

This Large White Ulster boar was bred by Mr William J McElroy of Rossdowney House, Londonderry, it was called 'Rossdowney Styles'.

"No man should be allowed to be President, who does not understand hogs", stated Harry Truman (1884–1972). He was the 33rd President of the United States of America and although it is highly questionable what, if anything, his successors have known about pigs, Harry Truman would have been conversant in matters porcine.

Before the Great War (1914–1918) he was a farmer and would probably have been able to identify America's top swine breeds such as the Duroc–Jersey, Poland China, Chester White and the Hampshire.

At the same time that Harry Truman was working the land of Missouri (ie the early 1900s), some members of Northern Ireland's pig fraternity were embarking on an exciting project. These people were all enthusiastic breeders of the Large White Ulster.

This breed, sadly now extinct, is believed to have evolved from the old 'Greyhound' pig of Ireland. An illustration of this ungainly beast, depicted by HD Richardson (*Domestic Pigs,* 1846), shows it to be bristly, coarsely fleshed, long-legged, raggedly tailed, and altogether lacking in refinement. However, crossing with

English breeds, and better feeding and housing, brought about progress. By the 1870s these improved pigs were becoming more widespread in Ulster.

In February 1905, the council of the Royal Ulster Agricultural Society gathered at 7 Donegal Square West, Belfast. At the meeting Mr Edward Coey moved the following resolution: that the Society should publish a herd book to be called the *Ulster Pig Herd Book* for the registration of the best sows and boars of the White breed. After discussion, it was agreed that the matter be referred to the 'Swine Committee'.

What unfolded over the next three years led to the announcement in January 1908 that the Society, in consultation with the Department of Agriculture, had agreed to a request from pig breeders and bacon curers to establish a Herd Book for the native breed. The designated name was the Large White Ulster and the breed Secretary would be Royal Ulster Agricultural Society's Manager 1897–1930, Mr Kenneth MacRae.

At that year's spring show at Balmoral, two classes were held for Large White Ulster pigs and those entered were reported as being "excellent specimens of the breed". Among the prize-winners at this historical event in the boar section were Mr J Cunningham, Belmont, Antrim ('Right Stamp of Belmont'); Mrs E Giffen, Springhill, Crumlin ('Young Revenge'); Thomas Lindsay, Derryboye House ('Ulster Jack') and Mrs Towley, Magherscouse, Ballygowan ('Pride of Erin'). Winning exhibitors in the female classes included Mr William R Smyth, Ballyalgin and Robert Suffern, Ballyclan House, Crumlin ('Ballyclan Polly').

After the First World War the membership of the Large White Ulster Pig Society continued to expand. By 1926 it had reached 102, made up of 86 members from Northern Ireland, six from Southern Ireland, four from England, four from Scotland and one from Wales. During the mid-1920s it was reported that a consignment of two boars and gilt had been sent to Italy.

In 1925/26 the 11th Herd Book for Large White Ulster Pigs was published. It contained pedigree details on 444 boars and 613 sows. One of the male entries was called 'Kilderry Forest Ruler', bred by Lieut–Colonel GV Hart, Kilderry, near Londonderry.

This pig was born on the 2 July 1925 out of a homebred sire and dam. At eight months of age he was entered at the Balmoral Spring Show and Sale where, after being awarded a prize, he went on to fetch 16 guineas, paid by the Large White Pig Society's Patron, His Grace the Duke of Abercorn. Much good blood was dispersed through the country by sales such as this.

The breed's representation at Agricultural Shows in the Province also served to generate interest. The Large White Ulster Pig Society provided 15 medals for competition through nine different agricultural societies. In the mid-1920s 12 gentlemen were listed by the Society as being qualified to judge Large White Ulster classes.

Alphabetically they were: Bell, Robert, Fruithill, Hillsborough, Co Down; Cunningham, J, Glencairn, Belfast; Galbraith, WH, JP, Craigadooes, St Johnston, Londonderry; Glenn, Robert, Glenvale, Campsie, Londonderry; Lindsay, Thomas, Derryboye House, Crossgar, Co Down; Low, George, The Farm, Ballywalter Park, Co Down; McElroy, William, J, Rossdowney House, Londonderry; Nelson, WR, Ardlauragh, Glenavy, Co Antrim; Roulston, Thomas, Ranelly, Omagh, Co Tyrone; Short, James, JP, Wood Park, Anney, Beragh, Co Antrim; and Wallace, John, Anticur, Dunloy, Co Antrim.

These men would have had a clear vision of the breed's official 'Standard Description'. It described the ideal pig, "Head – moderately long, wide between the ears; Ears – long, thin and inclined well over the face; Jowl – light; Neck – fairly long and muscular; Chest – wide and deep; Shoulders – not coarse, oblique, narrow plate; Legs – short, straight and well set, level with the outside of the body with flat bone not coarse; Back – long and level; Sides – very deep; Ribs – well sprung; Loin – broad; Quarters – long, wide and not drooping; Hams – large and well filled to the hocks; Tail – well set and not coarse; Coat – small quantity of silky hair, and Skin – fine and soft".

The latter aspects, regarding the breed's fine and soft skin, meant that it was well suited for use in the 'Ham and Roll' bacon which Ulster curers were shipping to the North of England and some industrial areas in Scotland. Because the Ulster White pigs bruised easily when transported live, they were often killed on-farm and the carcasses, when cool, taken by cart to the markets and bacon factories.

Once here, the carcasses would be cut in half and the hams (and sometimes the shoulders) split for separate curing. The skin was left on and post-curing the side would be rolled and tied with string in rings about one inch apart. The boned hams and shoulders were also tied up.

In 1936 it was projected that the yearly requirement for bacon pigs would be around 650,000. This was to be made up of 430,000 for the Ham and Roll trade, 50,000 for shipping and between 150,000 and 170,000 for the Wiltshire curers.

One of the Province's Wiltshire Bacon curers was that of Henry Denny and Sons Ltd, Obins Street, Portadown. In the mid 1930s the firm placed advertisements for pigs between 1 cwt, 2 qrs and 7 lbs and 2 cwt, 0 qrs and 13 lbs live weight. Furthermore, it was stated that Large

White Ulster or Ulster crosses were unsuitable because they laid down too much fat. The raw material for good Wiltshire curing was stated as being 'correctly fed pigs of the Large White York breed'.

During the late 1930s the call for pig breeders and feeders to 'Go York' came from the Ulster Pig Board and various County Agriculture committees. In some cases, however, it would seem that farmers were reluctant to switch. At a meeting of Armagh's County Agricultural committee, Mr McLaughlin seemed to question the Secretary's powers of persuasion when only two of the boar premiums went to Yorks and rest to pigs of the Large White Ulster breed. "I cannot force a pig down a man's throat," was the exasperated reply from the latter.

The majority of pig farmers were not hard to convince that the future lay with the York and they were supplied with good breeding stock from the Province's early herds, such as those of the Irish Peat Development Company, The Birches, Co Armagh; the Hillsborough Agriculture Research Institute; and the Ranfurly Stock Farm, Dungannon. These enterprises dispersed their bloodlines by holding on-farm sales, and participating with other breeders, at the RUAS show and sales.

At the Royal Ulster Agricultural Society's February show and sale in February 1936, there were 348 pigs put forward. Eighty-eight percent of them were Large White Yorks with the remaining twelve percent belonging to the Large White Ulster breed.

Throughout the 1940s and 1950s this trend away from the Ulster continued unabated. In January 1961 a farmer called Joseph Cooper of Warrenpoint was reported as having the last "full-bred Ulster pig alive".

The Rare Breeds Survival Trust was formed in the early 1970s and was able to offer much needed protection and encouragement to several of the native British pig breeds. Sadly, it was too late for the Province's native breed. When the inventory was completed, it was realised that the Large White Ulster had drifted into extinction.

> *"call for More Pigs ... More Pigs ... and More Pigs"*

Looking back to the days when they couldn't get enough pigs

As the call for More Pigs ... More Pigs ... and More Pigs went out from bacon companies during mid 1930s, many livestock owners throughout Ulster made the decision to make room on their farms for a pen or two of 'grunters'.

Having settled the matter regarding accommodation for the pigs, the next question may have concerned the method of feeding. Basically, there were three different systems; they could feed dry and provide water to drink separately, feed sloppy (adding 3 lb of water to 1 lb meal) or feed a semi-dry mash (1 lb meal to 1 lb water) with a little additional water given after feeding.

At this stage, having decided where to keep the pigs and what system of feeding to employ, the next perhaps more difficult question was, "Which Meal Company will I use?" At this time in the 1930s there were several reputable animal feedstuff businesses in Northern Ireland vying for trade.

In an extensive range produced in their mill at Donegal Quay, John Thomson and Sons Ltd were high grade weaning, rearing and fattening meals for pigs. Not only did the firm have a fully qualified chemist analysing various rations in their laboratory, but they also conducted

This pedigree Large White boar was bred in a County Down herd in the 1930s.

experimental work on Downshire Farm at Ballygrainey, County Down. 'Thompson's Pig Rations for Profitable Carcases' was the company's slogan used in 1936.

'Growlean' and 'Clarendo' were two more feedstuffs most worthy of consideration in the 1930s. The former product being 'scientifically balanced' and manufactured by James Irwin of Scotch Street, Thomas Street and Lr English Street, Armagh and, the latter by White Tomkins & Courage Ltd, Clarendon Mills, Belfast. Although best known for their 'White's Wafer Oats' and 'White's Jelly Crystals', this

company's 'Clarendo' was a wide spectrum 'fine food' used in the feeding of horses, cattle, sheep, poultry and pigs.

Having chosen the company from which to source their meal, farmers starting off in pigs at any time may have pondered over the use of pig powders such as 'Porkaline', which could be obtained from wholesalers Thos McMullan & Co Ltd, Beflast. Another tremendous product, this Porkaline was invaluable (used twice weekly) in the treatment of Worms, Cramps, Costiveness, Fits, Skin diseases and Rheumatism.

A mere snip at one shilling for thirty-two doses or six old pennies for fourteen doses, farmers using Porkaline could have been in the happy position of being able to send their pigs to market four weeks early.

In encouraging farmers to record the time taken for their pigs to reach slaughter weight and ensuing grading results the Ulster Pigs Marketing Board, in 1936, stated this information could then be used to identify superior breeding stock. That year the Board announced its intention to introduce a voluntary national pig recording scheme and litter testing station to spearhead improvement.

Speaking at a Board meeting that same year, Dr JS Gordon stated that the objective should be to have all pigs produced in Northern Ireland fed and killed here as opposed to being shipped out alive for finishing and processing across the water. He reiterated the Board's desire to increase the production of that type of pig the carcase from which would suit Wiltshire curing. This, Dr Gordon stated, would help gain a firm footing for Ulster bacon in the English market. To this end, he urged breeders during the following twelve months to buy a foundation stock that would enable them to produce the kind of bacon saleable at top prices. The Board, at this time, would 'assist' farmers by dropping the prices paid for heavy, fat pigs.

Following their decision to respond to the 'More pigs' call in 1936 and having decided on the matters around accommodation and meal supplier, the next step would have been to actually get some pigs. While the local market may have sufficed for those farmers wanting to fatten up a couple of batches, the Royal Ulster Agricultural Society's Spring and Autumn pedigree show and sales would have been the place to go for those people wanting something that little bit special.

Take show and sales held in September 1936, for example, where pedigree Large White pigs from most of our leading herds came to Balmoral including Ballywollen, Ranfurly, Fintona, Cumber and Wood Park. They had to arrive not later than 3pm on 23 September in order that Ministry of Agricultural officers could select those pedigree pigs to be granted a premium.

When the sale took place the following day, it was noted that among those gathered was Mr Alec Hobson, Secretary of the National Pig Breeders' Association, a well-known figure. Indeed the whole Hobson family were celebrated servants to agriculture. Alec had two brothers called Harry and George, the former being the famous auctioneer who founded Harry Hobson & Co, Southampton Row, London and the latter Secretary of the British Friesian Cattle Society between 1911 and 1946.

That such a high-profile figure among the pig breeding fraternity should be present a this sale was significant because, when the day's business was completed, those animals sold included a female, later reported as being the 'first breeding pig ever shipped from Northern Ireland to England'. Mr James Short of Wood Park, Anney, Beragh, County Tyrone had consigned this first prize-winning sow, 'Woodpark Molly Bane 3rd', which was purchased by Mr WW Ryman, The Manor Farm, Lichfield, Staffs.

A life member of the NPBA, Mr Ryman's large herd of pigs, registered under the prefix 'Wall', was held in high esteem. That same year, a breeding sow ('Wall Jubilee Maple') bred and exhibited by Mr Ryman won the Gold Medal at Royal Agricultural Society of England's annual show at Wolverhampton, while herd mates picked up twelve silver and two bronze medals at other agricultural shows staged up and down the country.

At the aforementioned sale, Mr Ryman paid thirty guineas for 'Woodpark Molly Bane 3rd', which was at the time of sale in pig to a 'Rossdowney' boar bred by Mr WJ McElroy of Rossdowney House, Londonderry. According to Volume 54 of the National Pig Breeders' Herd Book, having been taken across the Irish Sea 'Molly Bane 3rd' gave birth to twelve piglets on 27 October 1936.

It was the day's second top price, the best being Mr Brown's (Ranfurly, Dungannon) 'Ranfurly Boy 18th'. Details regarding the transfer of this pig and several others at the Balmoral Show and Sale held during September 1936 are printed in the table below.

J Short and Sons, Beragh – Woodpark Molly Bane 3rd to WW Rymen, Lichfield for 30 gns.	**Robert Brown, Ranfurly, Dungannon** – Ranfurly Boy 18th to Nelson for 20 gns.
Robert Brown, Ranfurly, Dungannon – Ranfurly Bandmaster 31 to A McDonald, Carrowadore for 14 gns.	**John C Crossle, Anahoe, Ballygawley** – Anahoe Bandmaster to W Gabbie, Ballywollen, Crossgar for 12 ½ gns.
Wm J Robb, Cumber House, Fintona – Fintona Prince to SD Reeve, Bangor for 12 ½ gns.	**Wm Gabbie, Ballywoollen, Crossgar** – Ballywoollen Bella 2nd to Drennan for 14 gns.
WJ Johnston, Beragh, Tattykeeran Violet to JM King, Brookborough for 12 guineas.	**Robert Brown, Ranfurly, Dungannon** – Ranfurly Bonetta 7th to TV Gibney, Banbridge for 14 gns.
Robert Brown, Ranfurly, Dungannon - Ranfurly Madam 8th to H Gray, Tardee, Kells for 14 ½ gns.	**Robert Brown, Ranfurly, Dungannon** – Ranfurly Madam 9th to WJ Stockdale, Downpatrick for 13 gns.
Leading prices from Autumn Pig Show and Sale held at Balmoral in September 1936	

"Just as it is today, one of the reasons that pig farmers bring their stock to the annual spring show at Balmoral is that a rosette or two will help them sell stock off the farm during the year."

Pigs in ascendency during post-war years

Even in those halcyon post-war days, some farmers may have been frustrated by 'red tape' as they filled in forms detailing the numbers of livestock on their farms. It had to be accurately done when applying for their quota of rationed feeding stuffs.

According to governmental newspaper notices published in May 1949, it was an offence (under Defence Regulations) for any farmers applying for rationed feeding stuffs, to overstate the numbers of livestock on their farm. Although the war was over, rationing, and form filling remained for a time.

After a hard day's work, one can just imagine farmers at that time doing the paperwork of an evening.

Before writing down the numbers of milk cows, their mind would have been transported to the byre-stalls. Scratching their heads, they may have been heard saying, "Now let me see, there's Biddy at the door, next to her is Doris, Sadie, Daisy, Young Biddy, Mabel, &&"

Although pig numbers had dramatically fallen during the war, by 1949 there was an estimated 365,000 in Ulster. The situation could have been described as a sow or two on most farms and these animals earned their ration by producing a couple of litters each year, the piglets of which could be sold off as weaners, stores or kept on the farm and fattened up for slaughter.

Many of these farms in the late 1940s would not have had enough females to justify the keeping of an adult male and so, as the need arose, sows and gilts were taken to 'visit the boar' on neighbouring farms where more pigs were kept.

The fact that many sow-owners across the Province would not regularly have seen fully grown adult male pigs did much to stimulate interest in when such animals were being exhibited at our agricultural shows. Take the 1949 Spring Show at Balmoral, when the Large White section comprised 23 junior gilts, 11 senior gilts, 14 sows, 22 junior boars and 8 senior boars.

Having gone through all the washing and brushing, the pigs entered in the aforementioned old boar class, would have been created something of a sensation. When in the ring, each animal would have been accompanied by two attendants clad in white coats. Their objectives would have been to ensure safe handling, keep potential combatants well apart by use of baton and board and present their boar before the judge in such as way as to extenuate the positive.

When judging the senior boar class at the 1949 Balmoral, Mr Vestergaad, owner of the Northants based 'Lyveden' herd, would have been assessing how each pig measured up to the breed's standard of excellence. Even in this class of heavyweights, he would have wanted to see a 'firm and free' action as they made their way round the ring.

Here are some details on the main contenders for Mr Vestergaad's top places: 'Spalding

This fine Large White boar was a champion in 1949.

Turk 15th', an eighteen-month-old English-bred pig owned by Mr Jack Campbell with four litter siblings having been selected for export to Russia; 'Dogleap Prince 10th', rising three-years-of-age and weighing over ten hundredweight (500 kgs), this boar had been bred by Miss Dorothy Robertson of Limavady and was exhibited by Mr David Moses and Sons, Seaskinore, Omagh; and fourteen month old 'Bardick Adept 152nd', born in a litter of seventeen, bred and exhibited by Mr JW Barbour of Castle Espie, Comber and the son of the RUAS male Champion in 1947 and 1948.

Three other pigs turned out in this class for 'Boar, born on or before 31 May 1948' at the 1949 Balmoral Spring Show were 'Dogleap Mainstay' (born 21 April 1948, bred and exhibited by Miss Dorothy Roberston, Dogleap, Limavady), 'Tring Turk 58th' (born 8 August 1947, bred by HW Bishop, Park Hill Farm, Herts and exhibited by Mr James Pollock & Son, Drumcannon Farm, Ballymoney) and 'Wall Gladiator 10th' (born July 5, 1947, bred by WW & WJ Ryman, The Manor Farm, Staffs and exhibited by Crossgar Poultry Services, Crossgar).

Having completed the judging of this class, Mr Vestergaard ranked the aforementioned boars in reverse order as follows: HC, 'Wall Gladiator 10th'; Res, 'Tring Turk 5th'; 4th place, 'Dogleap Mainstay'; 3rd place 'Bardick Adept 152nd'; 2nd place, 'Dogleap Prince 10th' and 1st place, 'Spalding Turk 15th'.

Just as it is a great honour for our present-day breeders to win a top place in the pig classes at Balmoral, so it was around fifty-eight years ago when the 1949 show was taking place. When Mr Jack Campbell saw the red rosette being awarded to his boar, 'Spalding Turk 15th', it would have been a proud moment but there was more to come. Having been place ahead of the top pig in the junior class (ie 'Inver Leader 27th',

bred by JC Welsh, Ailsa, Ballysnodd, Larne and exhibited by Irish Peat Development, Derryane, Dungannon), Mr Campbell's pig was made Overall Male Champion.

Considering that Mr Jack Campbell had only established his pedigree 'Carncullough' herd of Large White pigs with the purchase of 'Dogleap Primrose 13th' around three years before this 1949 show, this was a great start and 'Spalding Turk 15th' a worthy winner. The more knowledgeable spectators, watching the judging of the senior boar class at the aforementioned show, would have been familiar with the long-established herd in which 'Spalding Turk 15th' had been bred. Mr Alfred White of Spalding, Lincolnshire, had founded it in 1906.

At the time Mr Jack Campbell purchased his 'would-be' champion boar, Mr Alfred White's son, Mr Willis Martin White, was in charge of the large farm enterprise, the main occupation of which centred on the growing of daffodil and tulip bulbs. Three farms, all located in the Spalding area, were used and they were Little London Farm (350 acres), Pinchbeck Farm (650 acres) and Moultan Farm (400 acres).

The 'Spalding' herd of pedigree Large White pigs consisted of about 35–40 sows that were kept indoors but given frequent access to pasture. The objective was not to molly-coddle the pigs but to rear them up as naturally as possible. Mr White put great importance in selecting his breeding stock by eye, sticking to the standard of excellence laid down by the Large White breed standard.

For successive decades, many pigs bred in the Spalding herd won prizes at Agricultural shows throughout the country: In 1949, for example, Spalding-prefixed pigs won at shows organised by the Royal Agricultural Society of England, Essex Agricultural Society, Kent County Agricultural Society, Rutland Agricultural Society and Three Counties Agricultural Society

and the previously mentioned Balmoral Spring Show where Mr Jack Campbell's 'Spalding Turk 15th' took the top award.

Just as it is today, one of the reasons that pig farmers bring their stock to the annual spring show at Balmoral is that a rosette or two will help them sell stock off the farm during the year. Customers are always impressed when they are told that the pig they are looking at was closely related to a prizewinner. Such an animal is surely worth a pound or two more.

When the aforementioned Mr Jack Campbell took a fifteen-month-old daughter of the 'Spalding Turk 15th' to the Royal Ulster Agricultural Society's autumn pig show and sale, held on 19 and 20 October 1950, one can be sure people were told of her sire's show-ring credentials. His gilt, called 'Carncullagh Maple Leaf 7th', was placed first in a class of twenty-three and went on the win the National Pig Breeders' Association's Challenge Cup for best pig in the show. When one considers that there were over two hundred other Large Whites entered, this was a significant achievement.

"Here's our Champion", the auctioneer may have said as Mr Campbell's gilt came into the ring, adding, "Her sire won the MacRea Cup at this year's show, who'll start me off at 90 guineas … 90 guineas anyone – well, eighty then – seventy surely?"

Trade was brisk at that Balmoral Show and Sale in October 1950 and at the close of business ninety-eight boars and eighty gilts had cleared at an average of £40 7s 5d and £50 4s 1d, respectively.

How much did Mr Jack Campbell's, 'Carncullagh Maple Leaf 7th' make? Well, farmers opening their newspapers the next day would have had the answer in the headline "100 guineas for Champion Pig".

The bacon king of the early 60s

Who was the Royal Ulster Agricultural Society's Bacon Carcase King in the early 1960s? This would be a fine question for a rural quiz-night and it is one that, in attempting to retrieve an answer from their distant memories, some of *Farm Week*'s more 'senior' readers would be seen scratching their heads.

Well, before providing the answer to this conundrum, let's take a closer look at the Balmoral Bacon classes focussing in on the 92nd annual show, held 27, 28, 29 and 30 May 1959. At that event the wide-range of exhibits included one hundred and sixteen live pigs and sixty-seven dead ones. The former group comprised 97 Large White and 19 Landrace whereas those of the latter band were spread over two classes: Class 123 – 'Bacon Carcase, any pure breed, 135-160 lbs (dead weight), born on or after 2 October, 1958' and Class 124 – 'Bacon carcase, breeding to be stated, 135–160 (dead weight), 190–205 (live weight)'.

Those pigs entered for the 1959 Balmoral Show carcase competition had to be delivered 'on the hoof' to Denny's Bacon factory in Obins Street, Portadown between 20 April and 7 May. Following slaughter, dressing and curing, the pigs were delivered 'on the hook' to the

Mr McCutcheon's pedigree Landrace boar 'Kylestone Broson 20th', photographed at Newry Show in 1961 after taking the top prize.

Mr William McCutcheon won the Bacon Carcase Championship at Royal Ulster Agricultural Society's 92nd Balmoral Show held 27–30 May 1959.

Balmoral Showground for judging on 26 May. Undertaking this task that year was Mr THG Price of Messrs Henry Denny & Sons, Limited and Mr G McGuinness of the Pigs Marketing Board (NI), New Forge Lane, Belfast. All the measurements were taken by Mr Ivan Heaney M Agri of the Greenmount Litter Testing Station.

The Bacon carcase competitions provided an opportunity to raise standards within the Industry. Farmers visiting the show could spend time seeing how good feeding and breeding decisions could impinge on carcase quality in terms of length, back fat thickness, eye-muscle area and colour of fat. The Champion carcase on display in the hall during the 1959 Balmoral came from a Landrace pig, bred and exhibited by Mr William McCutcheon (c.1911–1970) of Ballyminetra, Bangor. His animals were registered in the *Landrace Pig Society's Herd Book* under the prefix 'Kylestone'.

Although Mr McCutcheon ran a dairy herd of around forty Friesian cows (with a couple of Jersey to thicken up the milk) his greatest interest and success was as a pig breeder. At a time when Ulster's pig scene was dominated by the Large White, Mr McCutcheon was one of pioneering breeders to work with the Landrace. This breed had been introduced to the Province in May 1954 with the arrival of fifteen pedigree pigs from Malmo, Sweden.

It was during that decade that Mr William McCutcheon travelled by car to the Lisnaskea, County Fermanagh, where he purchased his first Landrace pigs (two gilts and a boar) from Mr Reggie Allen's Lisoneil herd. The animals were barely more than weaners and so they were transported in the back of Mr McCutcheon's Ford Consul. The gilts cost over £100 each and the boar (called 'Lisoneil Bishop' 9269) £250 and it did not escape Mr McCutcheon that the total cost of his rear seat passengers exceeded the value of his car! These pigs proved worthy of the investment and got the 'Kylestone' herd of pedigree Landrace pigs off to a great start.

Although back in the 1950s and 60s there was less information available to pig farmers when it came to selecting breeding stock, those who spent time studying grading records and the like were best placed to make good decisions. Of course, it's was not and is not an exact science and perhaps many *Farm Week* readers will be able to cite animals that failed to meet their perceived potential or others that exceeded all expectations.

When the time came for Mr William McCutcheon to buy a new stock boar he studied information from several Scottish herds and then, by good fortune, chose to buy it from the 'Springfield' prefix located near Edinburgh. 'Springfield Broson 2nd' was purchased for £180. If an extra nought had been put on the asking price this boar would still have been

Mr William McCutcheon of Bangor with silverware won by pedigree Landrace pigs from his kylestone herd.

worth every penny! Not only did his progeny achieve top grades having been taken to West Belfast's Colin Glen bacon factory but those sold on for breeding realised high prices. 'Springfield Broson 2nd' was the sire of the aforementioned champion Balmoral bacon carcase back in 1959.

Visitors to William McCutcheon's farm in 1961 would have been able to see the five-year-old 'Springfield Broson 2nd' still very much 'on his legs'. At the time he was working alongside a son called 'Kylestone Broson 20th'. Although the senior boar had not been shown owing to Swine Fever restrictions being in place when he was at his best, the son had a good show career and had actually won that year's Balmoral Championship.

In view of this it is not surprising that when, some weeks later, Mr McCutcheon brought 'Kylestone Brosom 20th' to Antrim Show his hopes for another piece of silverware were high. While preparing the boar for the ring, however, a most unfortunate occurrence took place when, without warning, it ripped a gash

in Mr McCutcheon's leg with a tusk. After a visit to the local hospital where he received six stitches, the stalwart pig breeder returned to the showground and claimed the championship with the boar.

By winning top awards such as these, Mr McCutcheon had little difficulty in shifting pedigree stock. There was, however, another way that farmers could tap into the Kylestone strain and this was buy bringing their sows to the farm for 'service'. Mr McCutcheon's son, Howard, recently related how one such farmer was making his way through Bangor with his sow in a somewhat dilapidated trailer. It was when backed up in traffic along the seafront, that the sow made good her escape. As she exited the trailer to the sound of splintering timber, those tourists gathered along the promenade were treated to some additional light entertainment.

Not only did Mr William McCutcheon win the 1959 Balmoral Bacon Competition but he repeated the feat in 1960, 1961, 1962, 1964 and 1965. And so, here is the quiz question again, 'Who was the Balmoral Bacon Carcase King of the early 1960s?' Answer: Mr William McCutcheon of Ballyminetra, Bangor.

Delivering a welcome sup after a strenuous period calving

A bottle of 'Gaseous Fluid' or Black Mixture would have given a great start to any cow rearing a calf in the 1930s.

Most veterinary surgeons working in large animal practices will have, at some stage or another, had the experience of leaving a warm bed in the small hours to attend a difficult calving on a cold farm. Such an incident is portrayed in the following story.

During the mid 1930s a county Antrim practitioner received an urgent call from a farmer who was having great difficulty in calving one of his cows. Pulling on his breeches, the vet was understandably fed up. It was winter and glancing out his bedroom window he could see that there was a heavy fall of snow.

Once outside he cleared windscreen of his Ford Saloon, coaxed its engine into life and set off in the direction of Ballynure. The driving snow made the journey through winding lanes exceptionally difficult. The farm was situated on high ground and it was

This advertisemet for 'Gaseous Fluid' The Stock-owners Best Friend, was used during the 1930s.

The Stock-owners Best Friend

For Colic, Hoven or Blown and Internal Chills.

"*use* GASEOUS FLUID Regd
OR BLACK MIXTURE"

A World-famed Drink for Exhausted Animals especially after :—
Foaling—Calving—Lambing—Farrowing.
Price 24/- per dozen bottles.
How to treat "Cattle Ailments" 1/- post free.

DAY, SON & HEWITT LTD. 22, Dorset St. LONDON, W.1.

unfortunate that within half-a-mile of reaching it, a deep snowdrift brought the vet's journey to an abrupt halt. Going to the boot of his car, he filled his leather case with several calving hooks, a concealed embryotom and some bottles of drenches.

Throwing some calving ropes round his neck, he progressed the journey on foot. Half an hour later, tired and wet through, the vet reached the farm to be greeted with the words, "Where have you been all this time? I can't afford to lose this cow or her calf!"

Having been brought into a stone outhouse, lighted by a single Tilley Lantern, the vet saw the cow lying on a bed of straw. It was obvious she was in discomfort so he quickly removed his Burberry and shirt. Plunging his hand into a bucket of icy water, the vet groped about before finding a tiny piece of soap. Having lubricated an arm, he carried out an examination. It was a textbook case of 'head and ears only presenting'.

Combining great patience with the skilful use of ropes, hooks and a T handle, the vet managed to deliver a living calf, which was rubbed down with straw and placed at the head of its mother. "What about a drink?" to which the vet replied, "Oh yes, a cup of hot tea would be great, I'm absolutely famished". After a slight hesitation the farmer affirmed, "I'm talking about the cow, is it okay to give her a drink?"

At this point the vet may have overcome his embarrassment by reaching into his case for a bottle of Gaseous Fluid. In the 1930s this concoction, also known as Black mixture, was very good for reviving exhausted mothers after foaling, lambing, farrowing and calving. It was manufactured by the well-known firm established in 1833, Day, Son and Hewitt Ltd of 22 Dorset Street, London.

While the story concerning the vet attending a cow in the Ballynure district may not be entirely accurate, it is a fact that the aforementioned animal medicine company Day Son and Hewitt Ltd did have a stand at the Royal Ulster Agricultural Society's May Show at Balmoral in 1934. Their representative Mr JH Bond manned it.

Before expanding on several other uses of the company's 'Gaseous Fluid' and indeed another wonderful remedy known as 'Vetalenta', it may be of interest to reflect on this 1934 Balmoral. It stands out in the Royal Ulster Agricultural Society's history because it was at this show, opened on 29 May 1934, that the Kings Hall was officially opened by His Royal Highness the Duke of Gloucester (1900–1974). He was the third son of George V on whose order the new structure had been named 'The King's Hall'.

Building work on the new hall had commenced less than a year before in June 1933 and as such, all those involved were grafting to a tight schedule. Just two weeks before the official opening, the King's Hall had no floor! When the Duke of Gloucester was told of how much had taken place in so little time he was reported as having been astonished.

Back on the 'Stand 122' at this historic Balmoral show, aforementioned representative Mr JH Bond of Day Son and Hewitt could have been heard selling the company's original wooden Stockbreeders' Medicine Chests. They could have been seen on the most progressive farms across Great Britain and Ireland, purchased by those farmers wanting to keep their vet's visits to a minimum. Stamped with the company's name, these impressive boxes had several compartments for their range of drenches, dressings, powders and pills. Day Son and Hewitt were diligent in promoting their wares, and had a series of illustrated wall-charts dealing with veterinary matters. These were given away free to vets and farmers.

"The cow's udder is the money box of your herd", Mr Bond would have told those Ulster farmers visiting his stand, as he exhorted them to invest thirty-six shillings in a dozen bottles of 'Vetalenta'. This wondrous drink not only dealt with Hard Quarters and Swollen Bags but Udder Felon and Garnet as well. When expounding the virtues of 'Gaseous Fluid', as recommended by the County Antrim vet in the opening story, Mr Bond would have been at pains to point out that this was the 'Stockowners Best Friend' and in addition to being a valuable restorative following parturition, it could be used to treat stomach disorders such as Colic or Gripes and for relieving cases of Blown.

According to a Farmer and Stockbreeder advertisement published in the 1930s, Gaseous Fluid (or Black medicine) cost 24/- per dozen bottles. While that price may have been acceptable to farmers across the water, it wouldn't have gone down so well at the 1934 Balmoral show. We can be sure that when Day, Son and Hewitt Ltd's Mr Bond came up against the negotiating skills of Ulster farmers, it would have been a time for sharpening his pencil, dropping the price and throwing in a few bottles for free!

"Although today it would be highly inappropriate to put a dead hen through the postal service, back in 1905 it would have been a pragmatic course of action."

Straightforward veterinary advice that came for free

When describing the relationship between a veterinary surgeon and farmers, an old writer stated that they were inseparable, adding that 'the former was a hand-maiden to the latter'. Although this comment was made over one hundred years ago, it is one that the majority of *Farm Week* readers would fully endorse. It's a good thing when members of the veterinary profession serve members of the farming community to the mutual benefit of both parties. That said there is one thing better than 'straightforward veterinary advice' and that is 'straightforward veterinary advice that comes without charge'.

Take, for example, a case involving a County Armagh farmer who owned a rheumatic cow back in 1905. The animal was around ten years-of-age and, after losing her milk for about three weeks, was having great difficulty in getting up. Now, this farmer had a choice; he could call out his local vet and part with cash or he could lift his fountain pen and write to the *Irish Farmers' Gazette* newspaper, which had a 'free' veterinary advice column. Perhaps it's not surprising that the farmer took the latter option and, a week or two later, got the following advice in the newspaper. "Give the cow a purgative dose of Epsom or Glauber's salts night and morning in her food".

This cob is in good health but if he'd been a 'whistler' in the 19th century his owner could have sought help from the *Farmer's Gazette's* veterinary column.

If this heifer had been 'off colour' in the 19th century, a full dose of Glauber's Epsom salts may have proved of benefit.

Back in the 20th century it would seem that most livestock owners would have been well advised to keep a bag of Epsom Salts (Sulphate of Magnesia) or Glauber's Salts (Sodium Sulphate) near at hand for use, not only as a purgative, but to combat instances of 'internal inflammation' as well.

Take another example featured in the *Farmers' Gazette* which involved a cow with what seemed like a cold. She had been a bit 'off colour' at the time of purchase but her new owner thought it would pass. Having sought advice from the *Farmers' Gazette*, the cow's owner received the following instructions, "Give her a full dose of Epsom or Glauber's Salts. House her at night and give her steamed bran, moderately covered with treacle." Another farmer, writing to the Gazette over a century ago, relayed how one of his sows seemed rather inactive. A week later the newspaper's veterinary columnist advised, "Give her occasional doses of Caster oil or ... Epson salts"!

The advice given in relation to a yearling bull with scour was, however, a little more detailed. This animal had been on a diet of hay, turnips and, linseed meal and oatmeal served in milk gruel. On this occasion the owner was told that the method of feeding was at fault, recommending that the beast be given no more milk or gruel, but rather pulped mangols, linseed cake, barley meal and good, long hay. This was clear guidance but there was more to come, "Give him 1 dr sulphate of iron, 1 dr powdered calumba root, 10 gr powdered nux vomica, 1 gr arsenious acid and 40 br powdered liquorice".

In circumstances where the value of the sick animal did not, in terms of economics, justify calling out a veterinary surgeon, once again the *Farmers' Gazette* column provided a solution. This was admirably demonstrated in the next case involving an ailing turkey which had a swollen head and partially closed eyes. Having forward these details to the *Gazette*'s veterinary column, the Turkey's owner was advised to "keep it warm and administer 20 drops of whiskey in a spoonful of water twice daily"!

Another owner with a dead White Leghorn pullet on his hands parcelled her up and forwarded her to the Gazette's offices by post. Although today it would be highly inappropriate to put a dead hen through the postal service, back in 1905 it would have been a pragmatic course of action. As the mail was being opened that particular morning, the newspaper's clerk wouldn't have raised an eyebrow as they opened the package and marked it for the attention of the veterinary columnist. A week or so later, the post-mortem result and direction were published. "Your pullet died from an inflammation of the intestine. Give all the birds a pill containing 1 grain sulphate of iron and 1 ½ grains extract of gentian two times a day for a week. Also, dress the floor of the run with quicklime and dig over so as to destroy the ova of the worms that have passed through the birds".

Rheumatic cows, scouring bulls, ailing turkeys, dead hens and horses too were all brought under the spotlight through the *Farmers' Gazette* veterinary columns of one hundred years ago. A case involving the latter species was initiated when a horse-owner posed the question

'Does bolting produce a whistler?' Those unused to equine terminology would have struggled with this one but not the *Farmers' Gazette* Veterinary Advisor.

This person knew that the reference to 'bolting' concerned the rapid consumption of food and the term 'whistler' was used to describe that high pitched sound emitted by a horse with a problematic trachea. If the tone was low the animal was described as 'a roarer' but there was also 'grunters'. In one of the old veterinary books Professor Penberthy, attempting to throw light on the matter, stated that "whistlers and roarers do not necessarily grunt and all grunters do not necessarily whistle or roar".

When addressing the 'Does bolting produce a whistler' question, some background information was provided for readers of the 1905 column. Apparently the horse, a half bred colt, had been a good worker. Then one day he started to bolt his corn and some of it stuck in his throat. The vet was called and the obstruction was cleared. A week or two later the owner took the colt up from grass only to discover that 'He was a whistler'.

On receiving his reply from the veterinary column the colt's owner was told that the condition had not come from the fact that the animal had bolted its corn but rather because it had contracted a cold when at grass. It was further suggested that the owner would do well to apply a blister to the colt's throat and dampen his hay and oats before feeding.

Had a veterinary surgeon actually visited the aforementioned colt on its home farm in 1905, we can be fairly certain that the diagnoses and advice would not have been imparted in a few sentences as it had through the paper. For those farmers whose local vet was prone to giving long-winded advice, the column provided a refreshing option in certain circumstances as was illustrated in the following case of straight forward advice that came without charge.

Farmer: 'I have a cow, twelve to thirteen-years-old that calved towards the end of May in the year 1900 and was later put to the bull. She passed first twenty-one days without showing any signs of being in season. She turned out not in-calf. Later, she came in season again several times and was given a change of bulls. She has not retained service and has been a puzzle to farmers in my neighbourhood. I am certain she did not abort. I would be very grateful if you would suggest a remedy. She is a very good milker and gives about a gallon a day.'

Farmers' Gazette Veterinary Advisor: 'I would advise that you get her ready to meet the butcher'.

FROM THE DISPENSATORY (1899)

Sulphate of Iron – (Green Vitriol) A valuable remedy in anemia and general debility.

Sulphate of Copper – (Blue Vitriol) When used internally this will produce a tonic effect. A foot-rot ointment can be made by mixing with lard and Stockholm tar.

Nux Vomica – The seeds of the strychnos nux vomica, a tree common to the islands of the Indian Archipelago.

Gentian – A vegetable tonic used combined in the form of powder with Iron, ground ginger. Used to promote strength.

Belladonna – (Deadly Nightshade) In the form of extract proves one of the most valuable remedies in catarrh, pharyngitis, laryngitis, bronchitis and nervous excitability.

• A Blister could be applied as a counter-irritant in the hope that as the animal's natural body defenses rose to bring healing to the blistered area, the original and main malady would also be addressed. One of many blister recipes included: Powdered Cantharides (Spanish fly), Powdered Euphorbium, Oil of turpentine, Oil of Origanum and Hog's lard.

Extracts from *The Cattle Doctor* by George Armatage, MRCVS

"the good shepherd should 'have plenty of grit' about them to cope with what could be at times, difficult working conditions."

Heavy snows of 1937 claimed the lives of hundreds of sheep

When it came to the recruitment of a full-time Hill Shepherd in the 1930s, some of our larger landowners may have encountered difficulties, as an increasing number of young people were plumping for general farm work where they could experience the thrill of driving a tractor.

At the social gatherings of young farmers during this decade, as young men worked their way along tables laden with food, they would have been heard sharing tractor stories and debating the advantages of Spade Lugs over Cleats. For them, the smell of burning TVO fumes was a heady mix. With tractors on the most progressive farms, there would be no more trudging behind farm horses. Jobs such as ploughing and harrowing took on a whole new meaning when the power was being supplied from a spanking-new tractor, be it a Ferguson, Fordson or Case.

What some would describe as a 'light covering of snow' presented little problem for the flock on this County Antrim farm.

During this decade those young men and woman at Young Farmers meetings who still wanted to talk about steaming up ewes, lambing percentages and the most recent incidents of maggot fly were 'old school' and as such, in the minority. How could the thrill of roaring across the turf in a freshly painted tractor, compare with the dull, repetitive labour surrounding the lambing, dipping and clipping of sheep? No, during this new era of the 1930s it took a special person to take on the job of full-time shepherd.

Although most sheep across Northern Ireland would have been kept in small flocks on mixed farms, there were some large ones, such as that run on the Sperrins by Major CAM Alexander. Supposing the Major had been looking for a new shepherd back in the thirties, what qualities would he have been seeking in the ideal candidate? Well, a lack of interest in tractors would, undoubtedly, have been a distinct advantage.

For Major Alexander and his peers, the best type of shepherd to employ was someone who would 'take ownership' of the flock. They would refer to 'My Ewes' or 'My Lambs' and generally care for the animals as if they 'legally' owned them. With this type of attitude, the well-trained shepherd would have needed a minimum of supervision. A mild manner would have been essential in a good shepherd who, working alongside a close-mouthed dog, could handle the sheep without a lot of fuss.

When one of the old authors, Mr JFH Thomas, was commenting on shepherds in his book titled *Sheep*, he stated that "the best ones appreciated that these animals had a higher standard of intelligence than was usually accredited to their species". He also observed that, "although many shepherds were slow and reluctant in speech, the best of them had rich minds and having spent much of their work away from other men, and were usually great philosophers".

This aspect of 'working alone', would have presented the best shepherds with little difficulty, indeed, putting it into a Northern Ireland context, they would have been happier taking their sheep up into the peaceful hills than down to the bustling livestock marts conducted in East Belfast by companies such as Allams, Robsons and Colgans.

Finally, it was the opinion of the aforementioned author that the good shepherd should 'have plenty of grit' about them to cope with what could be at times, difficult working conditions. The fact that over the centuries many hill shepherds have lost their lives is well known. In one Scottish churchyard twelve shepherds were buried in one day following a terrible spell of weather.

Although no shepherds were reported as having lost their lives during the heavy snow storms that swept Northern Ireland during 1937, the same cannot be said about sheep. At 12 noon on a Thursday in mid-March that year, a heavy snow started to fall and continued uninterrupted until the early hours of Friday morning. This break, however, was temporary because around 12pm that same day it began again, this time continuing for twelve hours.

The situation was exacerbated by the biting east winds which created deep snow, especially in hollow areas across the countryside. Two trains travelling in opposite directions on the line near Aughnacloy were reported as having become buried in a huge snowdrift.

Roads, too, were severely affected with many people having to abandon their motors, especially across Antrim and Down. Finding his road blocked, one dairy farmer from the former county was reported as having used his barn door as a sleigh. Harnessed to a horse, it was used to convey milk cans to Carryduff, from where they were put on a train bound for Belfast. Unfortunately not all dairy farmers

showed the same initiative and the city had to satisfy demand by bringing milk in from Scotland and England.

Although rubber boot salesmen and young snowballing hooligans may have seen some merit in the persistent snow, the same cannot be said of flockowners, especially in the case of those with sheep on high ground. During these March 1937 blizzards hundreds of sheep were reported as having been lost on the Sperrin Mountains, Mourne Mountains and across North Tyrone.

Even in the case of sheep rescued from the snowdrifts, problems could follow with a condition known as 'Snow Fever', the symptoms of which were nasal and bronchial catarrh. According to information made available to Ulster farmers at this time, the sheep caught in drifts did not suffer from getting too cold but rather from over heating. Although blow-holes in the drifts may have allowed some fresh air to reach the sheep, in many cases the air surrounding them was stagnant. With sheep being especially sensitive to a build up of ammonia fumes, problems could ensue following rescue.

Those shepherds across Northern Ireland who, following the terrible conditions of March 1937, found themselves having to deal with cases of Snow Fever could have taken the following advice, which was given out at the time. It involved the administering of a stimulant, the recipe for which is printed below.

Recipe for a Stimulant to be given to sheep suffering from exposure that, in 1937, was found to be very suitable.

Liqueur acetate of ammonia	½ ounce
Sweet Nitre	½ ounce
Sulphate of Magnesia	2 drachms
Water	½ pint